# Water Re-Use and the Cities

**The University Press
of New England**

SPONSORING INSTITUTIONS
Brandeis University
Clark University
Dartmouth College
The University of New Hampshire
The University of Rhode Island
The University of Vermont

Edited by Roger E. Kasperson
and Jeanne X. Kasperson

# Water Re-Use & the Cities

Contributors: Daniel Dworkin, Stephen Feldman,

Roger E. Kasperson, David McCauley, Thomas J. Nieman,

John T. Reynolds, and John Sims

Published for Clark University Press by

The University Press of New England

Hanover, New Hampshire, 1977

# Contents

## Appendixes

# Tables

# Figures

# Preface

American cities are on the point of instituting new technological and managerial systems for the provision of water supply to their populations. In areas of
the Southwest this is the result of growing widespread regional imbalances
between demand and available water resources. In other cases water is
simply unavailable where it is needed, or droughts threaten the adequacy of
supply. Even in regions of relative plenty, cities are finding it increasingly
difficult to supply their growing populations and to satisfy the increased appe-
tites of those populations for water of high quality at reasonable prices. City
officials cast covetous eyes at more distant sources, but political and legal
obstacles thwart traditional solutions. Nature is no less troublesome: periodic
droughts require a margin of safety in all supply systems, thereby burdening
the consumers with added expense.

Water supply is but one element, of course, in the management of urban
water resources. The environmental movement in the United States has
spawned a rush of antipollution legislation, culminating in the Clean Water
Act Amendments of 1972 (P. L. 92–500). Extravagantly ambitious in scope,
the legislation commits the nation to a goal by 1985 of "zero discharge" of
pollutants into waterways. Implementation is lagging behind the legislative
timetable. In fact, 1975 arrived with only $1.034 billion of the Act's $18-billion
commitment actually spent and with only some 28,000 of an estimated
47,000 discharge permits actually in effect. Moreover, the language of the
permits all too often reflected existing treatment capabilities. This is not to
overemphasize the limited pace of progress to the far-reaching goals. Far
more significant is the impact that even limited progress will have on the
provision of water supply for the future American city.

By the mid-1970's conventional wisdom held that re-use of wastewater
must increasingly help to meet the future water needs of the American city. A
succession of national assessments has designated high priority to increased employment of water re-use. Despite some concern over unresolved uncertainties, the more conservative professional managerial and
engineering associations have endorsed this general view. Yet the obstacles
in the way of major changes remain poorly understood and threaten to hin-
der that progress so increasingly recognized as vital to the national interest.

Several problems have assumed prominence. In 1975 the health ramifications attendant on the re-use of wastewater for potable purposes remained a
troublesome issue characterized by a number of significant uncertainties. A
clear national effort to provide a comprehensive assessment of risks has yet
to be launched. In the absence of experience with re-use systems, the de-

velopment of methodology to provide the sorts of analyses and numbers critical to decision-makers is only beginning. Much debate has raged over likely public response to the whole notion of recycling wastewater, particularly for drinking. With the great fluoridation controversy still fresh in managerial memory, a comprehensive understanding of public assessment and response is essential. The results of fluoridation thus far have been ambiguous if not contradictory. Moreover, managers themselves may well act as the gatekeepers—those who will chiefly determine the entrance into public policy of this innovation. What role may they be expected to play in promoting or resisting the new technological and managerial systems demanded by increased water re-use?

The present book addresses these problems. As with most research, the finished product has not emerged in the orderly and rational way that prepared results tend to suggest. An initial preoccupation with the public acceptance issue as characterized by professional managerial journals led to a concern with the attitudes of public health officials, consulting engineers, and city officials. The resident microbiologist at Clark University, acting as an antidote to the buoyancy of social scientists, eventually succeeded in sensitizing the rest of the project members to the pervasive uncertainties connected with public health issues and their implications for research design and public policy. A grant from the Massachusetts Board of Higher Education permitted the addition of a praxis component focused upon community education. The ensuing discussions—open and often heated—with water-supply managers, state public health officials, city mayors, and citizen groups proved invaluable to an emerging understanding of the concerns of different actors in future water re-use innovations. It happened also that a new colleague at Clark University had conducted basic research on the national water re-use program in Israel. The study to follow, in short, grew by bits and pieces, but always in the direction of greater scope and increased policy concern.

The results of the research are frankly disquieting. The evidence betrays all too often a drift to new policies and institutions rather than informed progress and suggests a vacuum of national leadership. Moreover, it is clear that many of those who are responsible for leading others into new systems of urban water management will themselves be dragged kicking over the threshold by events beyond their control.

Multidisciplinary projects resting upon broad-based empirical research accumulate debts so numerous as to make acknowledgment in brief form inadequate. To the many who deserve the mention which is not noted herein, we beg your indulgence. We are truly grateful.

To the U.S. Office of Water Resources and Technology and the Massachusetts Board of Higher Education, we are indebted for the funding that supported much of the research cited. Clark University contributed support to coordinate the preparation of the manuscript as well as cartographic, computer and typing services. It also provided an environment in which multidisciplinary research receives more than token endorsement. The Hebrew

University Social Science Council contributed partial funding of Stephen Feldman's research on Israel.

Duane Baumann, an original project cosponsor, deserves special mention for his role as a principal investigator throughout the research; his contribution is present throughout the book. Gerald Karaska was very helpful during the troublesome final stages of the manuscript. At the beginning of the project, Marc Eichen, Russell Houde, Andrew Mills, and Walter Murphy participated in the formulation of concepts, review of the literature, and research design. James Johnson of the U.S. Army Corps of Engineers; James Basilico, Patrick Tobin, and Leland J. McCabe of the U.S. Environmental Protection Agency; Martin Cosgrove and James Matera of the Massachusetts Metropolitan District Commission; and Roger Rondeau of the Massachusetts Department of Public Health were all generous with their ideas and criticism.

For interviewing the general public in the field, we are indebted to Judith Dworkin, who conducted all interviews for the public acceptance study proper, and Alan Filo, Gerard Hyland, Judith Jussim, and Kay Sunday Xanthakos, who interviewed during the pilot study.

At our study sites a wide variety of people contributed generously of their time and knowledge. In particular we wish to acknowledge the help of James Phillips in Colorado Springs; Thomas L. Koderitz and John Williams in San Angelo; Ray Stoyer, E. W. Houser, and John Merrell, who contributed to our knowledge of Santee; Steve Fanning and Gordon Willis on Lubbock; and Franklin Dryden and John D. Parkhurst in Los Angeles County. In Israel, Peretz Darr supplied necessary unpublished materials, and Gil and Hella Yaneev aided in data collection.

Miriam Berberian and Terry Reynolds coordinated the preparation of the manuscript. Rene Baril, Pixie Monahan, and Cindy Wessler patiently typed the seemingly endless drafts.

Finally, we are indebted to our editor, David Horne, for correcting the numerous errors we overlooked and for supporting throughout the publication of the book.

Worcester, Massachusetts                                     R.E.K. & J.X.K.
                                                            February 1977

# Part I  Orientation & Background

# Water Re-Use:
# Need and Prospect
Roger E. Kasperson

**1**

America is a nation accustomed to abundance—and water. The average American already generates thirteen tons of waste each year, and this output is increasing. But just as the oil embargo rudely revealed that the profligate rate of energy consumption in the United States is not an innate right of American existence, so a new awareness is emerging that existing assumptions for water planning may not suffice for the future. Water withdrawals— those waters taken from the supply but not necessarily consumed—have increased steadily during this century, even if less dramatically than the growth of energy consumption. Between 1900 and 1970, while the population of the United States increased about two and one half times, total national water withdrawals registered a ninefold gain. In the two decades between 1950 and 1971 alone, the nation's daily use of water increased by two thirds.[1] As a result, urban water supply managers, confronting the rapid growth of demand caused by urbanization, industrialization, and changes in life style, have, for their sources of raw water, reluctantly cast their attention increasingly to distant high-quality water or to closer but more polluted rivers and lakes eschewed until now. National and local officials, as with energy, have not rushed to embrace a conservation ethic, to integrate new pricing systems, or to reconsider the time-tested methods of supply augmentation.

But it is clear that changes must come. The First National Assessment of the nation's water resources by the U.S. Water Resources Council in 1968 projected an increase of total water withdrawals from 269 billion gallons per day (bgd) in 1965 to 1368 bgd in 2020 and a rise in water consumption from 78 bgd to 157 bgd—an increase of 400 percent over the same period (Table 1.1).[2] Water for industry and electricity represents the bulk of this increase, but municipal uses were also expected to triple their withdrawal and more than quadruple their consumption rates. The final report of the National Water Commission presented a substantially similar picture of future water use needs. In projecting a number of alternative futures for water demand, the Commission recognized a number of essential factors, including (1) population, (2) the rate of national income growth, (3) per-capita energy consumption, (4) factors affecting demands for food and fiber for domestic use and for export, (5) government programs dealing with resource development and distribution, such as environmental protection goals and crop price support programs, (6) the rate of technological change, (7) recreational water uses, and (8) the price of water to the various users.[3] By altering the combinations of the four major variables of population growth, waste-heat disposal, dissolved oxygen in fresh water, and degree of sewage treatment, the

1. U.S. National Water Commission, *Water Policies for the Future* (Washington, D.C., Government Printing Office, 1973), pp. 6, 441.

2. U.S. Water Resources Council, *The Nation's Water Resources* (Washington, D.C., Government Printing Office, 1968), p. 4–1.

3. National Water Commission, (note 1, above), p. 11.

4. Water Resources
Council, pp. 1—3–5.

TABLE 1.1

*Estimated Water Use and Projected Requirements, United States
(million gallons daily)*

| Type of Use | Used | Projected Requirements | | |
|---|---|---|---|---|
| | 1965 | 1980 | 2000 | 2020 |
| | | *Withdrawals* | | |
| Rural domestic | 2,351 | 2,474 | 2,852 | 3,334 |
| Municipal (public-supplied) | 23,745 | 33,596 | 50,724 | 74,256 |
| Industrial (self-supplied) | 46,405 | 75,026 | 127,365 | 210,767 |
| Steam-electric power: | | | | |
| Fresh | 62,738 | 133,963 | 259,208 | 410,553 |
| Saline | 21,800 | 59,340 | 211,240 | 503,540 |
| Agriculture: | | | | |
| Irrigation | 110,852 | 135,852 | 149,824 | 160,978 |
| Livestock | 1,726 | 2,375 | 3,397 | 4,660 |
| Total | 269,617 | 442,626 | 804,610 | 1,368,088 |

Source: U.S. Water Resources Council, *The Nation's Water Resources* (Washington, D.C., Government Printing Office, 1968), p. 4–1.

commission concluded that total water withdrawals by 2020 would range from 570 bgd to 2280 bgd, while water consumption would range from 150 bgd to 250 bgd. Table 1.2 provides a comparison of the Water Resources Council and the National Water Commission projections.

Whereas the nation's demand (in the sense of requirements) for water is likely to increase over the next forty-five years by a factor of two to five for both withdrawals and consumption, the supply of water will remain relatively constant. The source of all water to the nation is precipitation, which averages thirty inches per year for the forty-eight contiguous states. The estimated mean annual runoff in the conterminous United States is approximately 1200 bgd; it is unlikely that this figure will change greatly over the next forty-five years.[4] Hence, the existing total national supply must somehow accommodate the increased demand for water.

To this point, however, the problem has been stated only in aggregate form, but the demand/supply balance does not assume an even geographic distribution, and this inequity forms the greater part of the national water supply problem. Annual precipitation ranges from less than four inches in areas of the Southwest to over 100 inches in coastal areas of the Pacific Northwest. A comparison of mean annual runoff with existing consumptive uses in 1970 (Table 1.3) revealed that the gross water quantity situation is already serious in the Rio Grande, Lower Colorado, and Great Basin regions. The Lower Colorado now consumes twice as much water as is available in 50 percent of the years, while the Great Basin and Rio Grande regions consume more than 60 percent of the mean annual natural supply. In a number of other regions, shortages for the entire area may be expected

5. Nathaniel Wollman and Gilbert W. Bonem, *The Outlook for Water: Quality, Quantity and National Growth* (Baltimore, The Johns Hopkins Press, for Resources for the Future, 1971), p.32.

| Used | Projected Requirements | | |
|---|---|---|---|
| 1965 | 1980 | 2000 | 2020 |
| *Consumptive Use* | | | |
| 1,636 | 1,792 | 2,102 | 2,481 |
| 5,244 | 10,581 | 16,478 | 24,643 |
| 3,764 | 6,126 | 10,011 | 15,619 |
| 659 | 1,685 | 4,552 | 8,002 |
| 157 | 498 | 2,022 | 5,183 |
| 64,696 | 81,559 | 89,964 | 96,919 |
| 1,626 | 2,177 | 3,077 | 4,238 |
| 77,782 | 104,418 | 128,206 | 157,085 |

periodically, and a projection of water needs indicates that these problems will only worsen in future years. Nathaniel Wollman and Gilbert Bonem argue that by 2020 extraordinary large-scale importation of water, shifts of population or industrial activity, or increases in supply through advances of technology will be required to meet the severe shortages in the Southwest and the region of the Upper Arkansas, White, and Red rivers.[5]

Even though the foregoing brief analysis reveals a number of serious supply/demand imbalances, it fails to uncover another pressing problem—the plight of the cities. It is the cities, with their population and industrial nodes, which are likely to spawn many of the severest future problems.

TABLE 1.2

*A Comparison of Water Demand Projections*

| | Projections of | | | | | |
|---|---|---|---|---|---|---|
| | Total Water Withdrawals (bgd) | | | Total Water Consumption (bgd) | | |
| | 1980 | 2000 | 2020 | 1980 | 2000 | 2020 |
| U.S. Water Resources Council | 443 | 805 | 1,368 | 104 | 128 | 157 |
| National Water Commission | — | — | 570–2,280 | — | — | 150–250 |

Sources: Water Resources Council, *The Nation's Water Resources*, p. 4–1. Water Commission, *Water Policies for the Future*, p. 13.

**Roger E. Kasperson**

6. Water Resources
Council, p. 4–1–4.

7. Ibid., p. 4–1.

TABLE 1.3

*Streamflow Compared with Current Withdrawals and Consumption*
*(billions gallons per day)*

| | | Annual Flow Available | | | | |
| Region | Mean Annual Run-Off[a] | 50% of the Years | 90% of the Years | 95% of the Years | Fresh Water Consumptive Use 1970[b] | Withdrawals 1970[b] |
|---|---|---|---|---|---|---|
| North Atlantic | 163 | 163 | 123 | 112 | 1.8 | 55 |
| South Atlantic-Gulf | 197 | 188 | 131 | 116 | 3.3 | 35 |
| Great Lakes | 63.2 | 61.4 | 46.3 | 42.4 | 1.2 | 39 |
| Ohio | 125 | 125 | 80 | 67.5 | .9 | 36 |
| Tennessee | 41.5 | 41.5 | 28.2 | 24.4 | .24 | 7.9 |
| Upper Mississippi | 64.6 | 64.6 | 36.4 | 28.5 | .8 | 16 |
| Lower Mississippi | 48.4 | 48.4 | 29.7 | 24.6 | 3.6 | 13 |
| Souris-Red-Rainy | 6.17 | 5.95 | 2.6 | 1.91 | .07 | .3 |
| Missouri | 54.1 | 53.7 | 29.9 | 23.9 | 12.0 | 24 |
| Arkansas-White-Red | 95.8 | 93.4 | 44.3 | 33.4 | 6.8 | 12 |
| Texas-Gulf | 39.1 | 37.5 | 15.8 | 11.4 | 6.2 | 21 |
| Rio Grande | 4.9 | 4.9 | 2.6 | 2.1 | 3.3 | 6.3 |
| Upper Colorado | 13.45 | 13.45 | 8.82 | 7.50 | 4.1 | 8.1 |
| Lower Colorado | 3.19 | 2.51 | 1.07 | 0.85 | 5.0 | 7.2 |
| Great Basin | 5.89 | 5.82 | 3.12 | 2.46 | 3.2 | 6.7 |
| Columbia-North Pacific | 210 | 210 | 154 | 138 | 11.0 | 30 |
| California | 65.1 | 64.1 | 33.8 | 25.6 | 22.0 | 48 |
| Coterminous United States | 1,201 | | | | 87 | 365 |
| Alaska | 580 | | | | .02 | .2 |
| Hawaii | 13.3 | | | | .8 | 2.7 |
| Puerto Rico | | | | | .17 | 3.0 |
| Total United States | 1,794 | | | | 88 | 371 |

[a]U.S. Water Resources Council, *The Nation's Water Resources* (Washington, D.C., 1968), p. 3–2–60.
[b]C. Richard Murray and E. Bodette Reeves, *Estimated Use of Water in the United States in 1970*, Geologic
Survey Circular 676 (Washington, D.C., 1972), p. 17.
Source: National Water Commission, *Water Policies for the Future,* (Washington, D.C., Government Printin
Office, 1973), p. 9.

## Water for the Cities

At first blush it would seem that cities should not have great difficulties with
water supply. In 1970 public water supplies accounted for only 7 percent of
total water withdrawals. The First National Assessment noted that in most of
the United States "ample water is usually available for municipal development within reasonable distances."[6] Moreover, water for municipal use enjoys high priority compared to other potential uses.

But the magnitude of the need is changing. Where 75 percent of the nation's population lived in metropolitan areas in 1970, by the year 2000 that
figure is expected to increase to 85 percent. Whereas in 1965 urban areas
consumed 5.2 bgd of water, by 2020 this figure is expected to reach 24.6
bgd, an increase of 7–16 percent in its share of total national consumption.[7]
For additional supplies many cities have already been forced to go outside

the metropolitan limits, and even outside their river basin. This problem has been particularly severe in arid and semi-arid areas but is also emerging in humid regions. Many cities are experiencing a pattern of diminishing returns of water coupled with increasing costs of development, and the resulting managerial squeeze points up the difficulties that pervade the search by cities for augmented supplies.

The cost of providing water in cities is therefore escalating rapidly. Capital and current expenditures in 38 standard metropolitan statistical areas in 1969 for water supply and waste collection and disposal were $30.50 per capita, comprising 20 percent of total municipal capital and 4 percent of other current outlays.[8] In addition, fragmented political and water utility districts hamstring efficient provision of water services and thwart economies of scale for urban residents. Moreover, the propensity of water managers to ignore actions that would conserve water and decrease demand complicates these problems. A study of the 1961-65 drought experience in forty-eight Massachusetts communities, for example, revealed that the characteristic response to stress was to search for new sources of supply augmentation.[9] Rarely did a community adopt or even consider adjustments directed toward decreasing use.

Considerations of water quality will not make urban solutions for new water sources easy. Many urban centers already provide water of questionable quality to their residents, particularly where the sources of water supply are heavily polluted. The seriousness of the problem is currently under review by the Environmental Protection Agency (EPA). Poor water quality in the streams in the vicinity of large metropolitan areas detracts from the value of this option for supply augmentation. Yet the lack of many new water supplies indicates that many cities increasingly must resort to poorer quality waters and re-use water that contains significant portions of effluent and contaminants—a situation not anticipated when current drinking water standards were promulgated. Meeting the increased demands for water in the nation's cities while simultaneously retaining a quality of water that will not be detrimental to the general health is a problem of first magnitude. The dilemma may well explain why all projections of future water-supply needs suggest that increased water re-use must play a central role in future urban water resources planning.

8. National Water Commission, pp. 441–442.

9. Clifford W. Russell, David G. Arey, and Robert W. Kates, *Drought and Urban Water Supply* (Baltimore, The Johns Hopkins Press, for Resources for the Future, 1970), pp. 70–81.

## Managerial Options for Meeting Future Water Supply Needs

Confronted with either current or projected shortages of water supply, a manager has a finite number of possible responses: he can decide to bear the shortage and do nothing, initiate efforts to make more efficient use of the existing supply, or, by various means, to augment the total available supply. Table 1.4 summarizes these options.

**Roger E. Kasperson**

10. Ibid., p. 189.          TABLE 1.4

*Managerial Options for Meeting Future Water Supply Needs*

A. *Bear the shortage*
B. *Make better use of existing supplies*
   1. Reduce demand
      (a) education and persuasion
      (b) metering
      (c) incremental cost pricing
      (d) restrictions
   2. Reduce waste and losses
      (a) reduce reservoir evaporation
      (b) eliminate leaks
      (c) water-saving devices
C. *Increase supplies*
   1. New surface water
   2. Ground water
   3. Better land management
   4. Cloud-seeding
   5. Desalinization
   6. Water re-use

*Bear the shortage.* One option, in the face of a growing unfavorable water-use/safe-yield ratio, is nonresponse—decision either to reduce demand or enlarge supply. Such a course of action, if sustained long, could imperil the general well-being of the population, to say nothing of the manager's professional security. In fact, the one unforgivable error for a water-supply manager—the one that carries the greatest risk—is to run out of water. Consequently, many managers overbuild, in terms of economic efficiency. The Massachusetts drought study suggests that most managers initiate action at the point at which the water-use/safe-yield ratio falls to 1.17.[10] Bearing the shortage is not a credible option in the eyes of either managers or their consumers.

*Making better use of existing supplies.* As Americans are discovering in many areas of resource management, they have been an extraordinarily wasteful and extravagant civilization. American cities can significantly improve their water-supply potential simply by re-examining carefully the possibilities for squeezing much more mileage out of existing systems. These means of coping include two major types of strategies—reducing demand and reducing waste.

The means of reducing demand include programs of public education and persuasion, metering, incremental cost-pricing, and restrictions. Although many cities attempt to *educate the public* regarding the difficulties of the water-supply situation in an effort to achieve voluntary reduction in demand, there is little evidence to suggest that such explorations lead to any sustained declines in water use. Installations of *meters*, by contrast, can have a significant impact on demand. Despite the fact that a number of major cities do not employ meters, this innovation alone has achieved reductions in water use ranging from 20 to 50 percent.[11] Steve Hanke found that metering in Boulder, Colorado, reduced water consumption by 36 percent by reduction of water use, increased attention to water leakage, and reduction of yard sprinkling.[12] Since metering ordinarily entails a shift from flat-rate to incremental pricing, it is difficult to pinpoint its effect. But metering clearly contributes to greater public sensitivity to and understanding of water-use patterns and costs.

Of the various means for reducing demand, the use of *incremental cost-pricing* has the greatest potential. Incremental (or marginal) cost-pricing refers to the principle that price should equal the cost of the last unit supplied. Customarily, cities employ block-rate pricing, which encourages high water use, favors large users, and prices the last gallon of flow, often the most costly, as the cheapest. Some metropolitan areas even utilize a "declining block-rate," which allows suburban residents to pay lower prices for lawn sprinkling at peak demand periods. Considerable efficiency can be gained by varying price according to such factors as peak-demand periods, distance from the source of supply, and elevation above pumping level. A pricing system that raised prices during peak demand periods could spread out the building of new additions to the water supply, thereby reducing the total cost of water to urban users and relieving some of the pressure upon existing supplies.

Finally, it is possible to impose *restrictions on use*, usually as a temporary adjustment. Water restrictions are usually the product of "crisis" situations and can result in a substantial leveling of peak-load demands, but they also breed public hostility toward the water managers.

A manager may carefully re-examine his storage and distribution as well as consumer use practices for prospective ways to eliminate waste or losses. *Reducing evaporation* in reservoirs and conduits could, in areas such as the Southwest, result in substantial water saving. Covering water surfaces and underground reservoirs are two costly alternatives. A design of reservoirs to achieve a lower area-to-surface volume is another means, which may have greater practicality. *Reducing leakages and seepage* within urban water distribution systems is still another method. Leakages not uncommonly account for 10–20 percent of total water withdrawal in a municipal system[13] and are as high as 35 percent in poorly managed systems. In cities with mounting costs for new water supplies, expenditures to reduce loss from the distribution system may be economically advantageous. Finally, cities in the future should consider educational programs and financial incentives to encourage users to install water-saving devices. The final report by the National Water Commission identified toilet flushing and lawn watering as two residential uses which appeared to be excessive and possibly targets for reducing demand.[14]

If water supply augmentation does, however, prove necessary or more economical, a number of options—including tapping new surface or ground water supplies, better land management, cloud-seeding, desalinization, and water re-use—are available for the manager's consideration. Since most cities have already exploited most of the readily available high-quality rivers

11. Peter W. Whitford, *Forecasting Demand for Urban Water Supply*, Report EEP-36 (Palo Alto, California, Stanford University, 1970).

12. Steve Hanke, "Demand for Water under Dynamic Conditions," *Water Resources Research,* 6 (1970), 1253–61. See also Charles W. Howe and F. P. Linaweaver, Jr., "The Impact of Price on Residential Water Demand and Its Relation to Systems Design and Price Structure," *Water Resources Research*, 3 (1967), 13–32.

13. Water Resources Council, p. 5–1–10.

14. National Water Commission, p. 14.

15. Ibid., pp. 230–231, 243.

16. Ibid., p. 357.

17. U.S. Office of Saline Water, *Saline Water Conversion Summary Report, 1971–1972* (Washington, D.C., Government Printing Office, 1972), p. 45.

18. National Water Commission, p. 336.

19. William H. Bruvold, "Public Opinion and Knowledge Concerning New Water Sources in California," *Water Resources Research*, 8 (1972), 1145–50.

and lakes, *new surface water supplies* are becoming much scarcer. Increasingly, significant new surface sources will involve long-distance transport, frequently involving interbasin transfers of water. Although this has been a preferred method of increasing supply among water managers and consulting engineering firms (the Feather River and Central Arizona projects in the West and the Cannonsville Reservoir for New York and Connecticut River water for Boston in the East are examples), legal and political controversy increasingly pervades such long-distance transfers. Moreover, the additional problems created by solving water-supply problems through massive engineering projects, coupled with mounting wastewater disposal difficulties, make water re-use an eminently more sensible long-term approach.

*Ground water* has also been a preferred source of additional water supply because of its generally high quality and low production costs. Ground water currently supplies about 22 percent of water withdrawals in the nation, much of it for agricultural irrigation, and this percentage is likely to increase somewhat in the future. For most major cities, however, ground water will not be a significant new source of water supply. In fact, there is evidence that ground water supplies are deteriorating in quality, and remedial actions may prove necessary to safeguard existing supplies.[15]

*Better land management practices* can also contribute to the augmentation of available water supply. Such methods as vegetation management in forest areas and river banks, snowpack management in forest and alpine areas, and treatment of soil surface to maximize runoff collection have an estimated potential valued at nine million acre-feet annually for the forty-eight contiguous states.[16]

*Cloud-seeding* is emerging as an alternative of considerable potential. The development of mathematical models to simulate the impact of cloud seeding upon precipitation has increased scientific understanding and predictability. Areas of orographic (mountain-induced) precipitation appear to offer the best prospects. Unfortunately the relationship between augmentation and supply is poorly understood. Moreover, the difficult issues surrounding ecological impact and legal implications rule out for the short term any major use of this source for cities.

*Desalinization of sea water* is another option that has made rapid strides over the past several decades. In 1972, prior to the escalation in the price of energy, long-range projected desalting costs stood at 25 to 35 cents per 1000 gallons at the plant for large-scale sea water systems and less than that for brackish water plants.[17] By 1970 some 33 plants were operating in the world, with an average capacity of 1.8 mgd.[18] Interestingly enough, there is also evidence that the public rates desalinized water, unlike renovated wastewater, very high in its preferences for new supply sources,[19] perhaps because of the inorganic rather than organic impurities in the untreated water. High energy and transmission costs suggest, however, that desalinized water will be a competitive alternative only in coastal cities short of other supply options.

Finally, within the supply-demand balance, *water re-use* undoubtedly will come into its own. The feasibility of water re-use is the concern of the present volume.

## Water Re-Use Defined

Water-supply managers, wastewater engineers, government officials, politicians, and the media have all proposed water re-use schemes (1) as
emergency measures for water supply, (2) as long-term solutions, (3) as
individual components for specialized types of need, or (4) as fringe benefits
of pollution abatement. So much confusion and ambiguity have characterized the discussion that some initial definition and conceptual clarification
are in order, both for the discussion in this chapter and for consistency in
appraisal throughout the book.[20]

*Wastewater* is any water derived from one or more previous uses. It usually has undergone some degradation in quality as a result of use. It may be
furnished untreated or collected and treated for some additional water use.
Treatment by one or several processes renders the water suitable either for
discharge or for re-use.

*Water reclamation* refers to an upgrading of the quality of the water to
make it usable again. If the treatment of wastewater restores it to the level of
quality which prevailed at intake into the system or renders the wastewater
largely free of pollutants so that such high-order uses as body contact or
ingestion may occur, then *water renovation* has occurred. Renovated water,
then, is best conceived as a subset of reclaimed water.

Although *water re-use* refers to the use of wastewater or reclaimed water,
confusion abounds even at this high level of generality. The exact nature of
re-use—whether planned or inadvertent, direct or indirect, or for potable or
nonpotable purposes—remains unclear. The treatment, if any, of wastewater and the method or manner in which the wastewater reaches the next user
are critical to the feasibility and desirability of re-use. *Inadvertent re-use* occurs when water is withdrawn, used by one party, returned to a water source
(for example, river, lake, aquifer) without specific intention or planning to
provide the water for the use of other parties, but is nevertheless so used.
*Planned re-use*, by contrast, involves the collection and purposeful provision
of wastewater for subsequent use. Water reclamation may or may not have
occurred. Intention and planning are thus central in this distinction.

Throughout the present book the terms *direct re-use* and *indirect re-use*
are subclasses of planned re-use. *Direct re-use* involves the transmission of
wastewater, untreated or treated, directly to some specific intended use. No
natural buffers, such as lakes, rivers, or ground water, serve as inter-
mediaries to expose the wastewater to natural purification and dilution systems. The variety of intended uses is large and has been employed over
many years in the United States. Irrigation for agriculture, parks, golf
courses, lawns, and greenbelts; processing and cooling water for industry;
water and nutrients for European fish farms; and water for such recreational
uses as fishing, boating, swimming, and scenic man-made water bodies—
the range of possible uses is wide. *Indirect re-use*, by contrast, returns the
wastewater to the natural hydrologic cycle, thereby capitalizing upon natural
disinfection by sunlight, sedimentation, and natural biochemical degradation
and dilution. Since this type of re-use involves natural buffers as an inter-
mediary stage, it involves greater temporal and spatial separation of treat-

20. For discussions of
these distinctions, albeit
ones that differ from
those presented here,
see George E. Symons,
"Water Reuse—What
Do We Mean?," *Water
and Wastes Engineer-
ing*, 5 (1968), 40–43;
Henry J. Graeser, "Wa-
ter Reuse: Resource of
the Future," *Journal of
the American Water
Works Association*, 66
(1974), 576–577.

21.   K. D. Linstedt, K. J. Miller, and E. R. Bennett, "Metropolitan Successive Use of Available Water," *Journal of the American Water Works Association*, 61 (1969), 610–615.

ment and re-use. There are a number of advantages. Greater time, for example, provides opportunity for monitoring and testing water quality before use, and spatial separation allows for dilution and improvements in aesthetic properties of the water. In fact, the greater the separations, the greater the advantage. Ground water recharge and upstream disposal are perhaps the most familiar and widespread of indirect re-use systems.

Reclaimed water can serve as a source of *potable* water or water for *nonpotable* uses, and it can go directly to the user or indirectly to recharge an aquifer, which in turn serves as a source of supply. The end use, potable or nonpotable, and the method of distribution, direct or indirect, each imposes different requirements on the physical design of the system and the safeguards to protect human health and well-being. Figure 1.1 diagrams several of the major types of re-use systems defined above. These definitions may be employed to contrast water re-use systems.

Although there are in the United States at present no municipal systems making direct, potable use of renovated water, Chanute, Kansas, and Ottumwa, Iowa, have used reclaimed water during brief emergency situations. And on a more permanent basis, Windhoek, Namibia (formerly South West Africa), has employed renovated wastewater. Also, an ambitious water re-use scheme is under way in Israel (Chapter 6).

Chanute, Kansas, processed and returned the effluent, after a seventeen-day lag time in a holding pond, to the distribution system. The resultant water quality was poor, however, and it is unclear exactly how many people actually drank the water. Ottumwa used water that was one third to one half effluent from the city of Des Moines for a similar period. Windhoek produced reservoirs of effluent, which were then mixed with water from conventional sources to provide a municipal supply. The important distinction between these cases and Windhoek (over and above the superior technology employed at Windhoek) was the latter's longer delay between producing and using the renovated effluent, thus allowing more extended exposure to the natural elements. Denver, Colorado, is currently investigating "potable reuse with the eventual goal of a nearby closed-loop system in the late 1990's" through a three-stage project which commences with a one-mgd demonstration plant in 1977. Eventually the objective is a 100-mgd plant, which would be more economically efficient.[21] Chapter 2 discusses these examples at greater length.

In the United States the most extensive indirect re-use of water for potable supply is in Whittier Narrows, California, which applies 50 mgd of treated effluent to gravel beds, charging areas for an aquifer. The effluent stored as ground water is subsequently pumped up for both irrigation and general municipal use.

Most re-use of water is for nonpotable purposes supplied directly from the treatment plant to the user. This category includes most of the present uses for irrigation, industry, and recreation. Some of the proposed innovative uses include dual distribution systems for cities in which separate supply lines would furnish potable water for drinking, cooking, bathing, and laundry, while reused water would be furnished for toilet flushing and for irrigation of lawns and gardens.

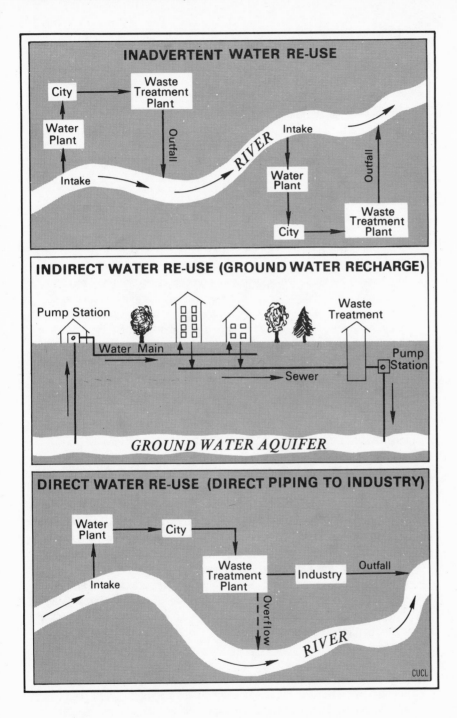

Figure 1.1: Types of Water Re-Use. After George E. Symons, "Water Reuse—What Do We Mean?" *Water and Wastes Engineering*, 5 (June 1968), 40–41.

22. Curtis J. Schmidt and Ernest V. Clements, III, *Demonstrated Technology and Research Needs of Municipal Wastewater* (Cincinnati, Ohio, Environmental Protection Agency, 1975), p. 119.

23. Curtis J. Schmidt, R. F. Beardsley, and E. V. Clements, III, "A Survey of Industrial Use of Municipal Wastewater," in *Complete Water Reuse: Industry's Opportunity*, ed. L. K. Cecil, American Institute of Chemical Engineering, Proceedings of the National Conference on Complete Water Reuse, April 1973 (New York, AICE, 1973).

24. Jerome Gavis, *Wastewater Reuse*, (Springfield, Virginia, National Technical Information Service, 1971), p. 1.

25. Louis Koenig, *Studies Relating to Market Projections for Advanced Waste Treatment*, Department of the Interior, Publication WP-20-AWTR-17 (Washington, 1966).

Nonpotable re-use of water usually does not require the extra treatment obtained by allowing the effluent to pass through layers of earth. Some places do, however, re-use nonpotable water indirectly. One previously mentioned is Whittier Narrows, which employs the water for both potable and nonpotable uses. Lubbock, Texas, boasts an innovative plan to apply wastewater directly to the land for irrigation and then to pump the underlying water table to supply a series of recreational lakes. The project, funded in part by the Bureau of Outdoor Recreation and the Department of Health, Education, and Welfare, was nearing completion in 1976 (Chapter 3).

## Water Re-use for Municipal Supply

Water re-use is a major element of national water supply. A comprehensive survey of re-use practices in the United States commissioned by the EPA identified some 358 municipalities that practiced some form of re-use.[22] Of these some 338 (94 percent) were involved with irrigation uses, mostly small-scale operations with crops directed to nonpotable or indirectly potable uses. An earlier 1971 survey had found a total of 143 bg (billion gallons) of water re-use, divided among 71.6 bg (50 percent) for irrigation, 37 bg (37 percent) for industry, 15.5 bg (11 percent) for ground water recharge, and 2.4 bg for recreation.[23]

With pressures of growing demand upon extant water supplies, many cities have resorted to polluted sources for supply. Indeed, one third of the total population in the United States already uses water containing one gallon of wastewater for every 30 gallons of flow. In many cases the wastewater has emerged only hours before from some upstream municipal or industrial sewer. In some instances the ratio of previously used water during low-flow periods is as high as one fifth of the total.[24] Louis Koenig suggests that as much as one half of the water withdrawn by the average municipality has been previously used upstream, and for cities in the upper 10 percent re-use bracket, the supplies have been used at least three times previously. Koenig reported the situation for many American cities as follows: Jefferson City, Missouri, had a re-use factor of 5.7 percent at the defined low flow and 1.8 percent at an average flow; Kansas City, Missouri, 6.6 percent and 2.4 percent, respectively; Omaha, Nebraska, 2.3 percent and one percent. Other data for St. Louis suggest comparable figures (excluding re-used cooling water) of 5.6 percent and 1.6 percent.[25] Obviously a great deal of re-use occurs.

*Risks in inadvertent re-use.* Unfortunately, present re-use is largely the inadvertent type. Because it is unplanned, the use of these polluted waters, even with modern treatment technology, involves a number of serious risks:

1. A breakdown in water treatment facilities may serve to carry contaminated water to users in great number. If dam waters are polluted, treatment facilities must be adequate in capacity, properly operated, with highly qualified chemists and bacteriologists in close

supervision, to ensure that breakdowns in the treatment barrier do not occur.

2. Even where conventional treatment is provided, the fate of viruses, particularly those of infectious hepatitis, is uncertain. Only a few virus particles need be ingested for infection to result. Most who are infected may not become visibly ill, but subclinical infections cannot be considered innocuous; their effects may be delayed or camouflaged among other illnesses, or the disease may be passed on to other members of the population.

3. Hundreds of new chemical compounds are being introduced into our environment daily. The likelihood of ingesting them increases greatly when contaminated waters are used as sources for municipal supply, because conventional water treatment is ineffective in removing them. Some of the chemicals, either alone or with others, have been shown to cause cancer, genetic damage, or birth malformations. Because the effects of such chemicals ingested in low concentrations over long periods are insidious and are likely to be similar to those manifested by aging, their significance is hard to establish.[26]

From 1961 to 1970 there were thirty-five known outbreaks of waterborne disease caused by contaminated water from public systems and ninety-three reported outbreaks caused by private water systems, resulting in more than 46,000 cases of illness and twenty deaths.[27] These risks have added to an emerging unease over the adequacy of present drinking water standards and the safety of municipal water supplies in the United States. The HEW Report of 1967 contended that 50 million Americans drink water that does not meet U.S. Public Health standards and another 45 million drink water not tested by the Public Health Service.[28] The 1970 Community Water Supply Study of 969 communities revealed that only 59 percent of these communities produced drinking water acceptable under its current recommended standards, only 10 percent had bacteriological surveillance programs that met the criteria, 90 percent either did not collect sufficient samples or collected samples that showed poor bacterial quality, and 61 percent had operators with inadequate water-treatment training.[29] A General Accounting Office Study of 446 American communities in 1972 found that eighty-one systems were delivering water whose bacterial content exceeded the limits of federal drinking water standards for two or more months and forty-four additional systems delivered water whose bacterial count exceeded the standards for one month.[30] In 1974 research linking possible carcinogenic agents in the blood of New Orleans residents to the city's water supply sparked a new comprehensive study of the safety of municipal drinking-water supplies.[31] The Safe Drinking Water Act of 1974 is, of course, directed to these concerns.

There is growing recognition, therefore, that although the nation is not running out of water, there is indeed a scarcity of high-quality water accessible to areas of need. It is abundantly clear that, given even partial achievement of the objectives of pollution abatement in the 1972 amendments (PL 92-500) to the Clean Water Act of 1965, the reclamation of wastewater and

26. Daniel A. Okun, "Planning for Water Reuse," *Journal of the American Water Works Association*, 65 (1973), 618.

27. Comptroller General of the United States, *Improved Federal and State Programs Needed to Insure the Purity and Safety of Drinking Water in the United States* (Washington, D.C., 1973), p. 7.

28. Department of Health, Education, and Welfare, *A Strategy for a Liveable Environment* (Washington, D.C., Government Printing Office, 1967).

29. Public Health Service, *Community Water Supply Study* (Washington, D.C., 1970).

30. Comptroller General of the United States (note 27 above).

31. Jean Marx, "Drinking Water: Another Source of Carcinogens?," *Science*, 186 (November 20, 1974), 809–811.

32. U.S. Congress, Senate Select Committee on National Water Resources, *Water Resources Activities in the United States* (Washington, D.C., Government Printing Office, 1960).

33. *Water Resources Planning Act of 1965*, Public Law 85–90, Sec. 102, 89th Cong., 1st Sess., 1965. See also Water Resources Council, "Principles and Standards for Planning Water and Related Land Resources," *Federal Register*, 38, No. 175 (September 10, 1973), 778–869.

34. Water Resources Council, (note 2, above), p. 1–23.

35. National Water Commission, (note 1, above), p. 311.

36. Duane D. Baumann and Daniel Dworkin, *Planning for Water Reuse* (Indianapolis, Holcomb Research Institute, Butler University, 1975), pp. 10–11.

its re-use will rapidly become an increasingly attractive alternative to long-distance transport of water for American cities.

At the federal level, groups like the Senate Select Committee on National Water Resources have called attention to the future role of re-use since the early 1960's.[32] The Water Resources Planning Act of 1965 requires that re-use be considered as one of the alternative methods of meeting future demands for water.[33] The National Water Commission Act of 1968 instructed the Commission specifically to consider water re-use as part of its review of present and anticipated national water resource problems.[34]

The U.S. Water Resources Council recognized the realities of re-use: "The large withdrawals estimated for 2020 in relation to run-off indicate that even with increased in-plant recycling a large increase in re-use would be required."[34] The final report of the National Water Commission provided several plans for meeting future national water needs through 2020 by more extensive planned re-use:

(a) By 1980, continuous re-use by manufacturers of only 20 percent of the projected total of 73 billion gallons per day (bgd), of municipal and manufacturing effluents would completely satisfy the projected 1965–1980 increase in water withdrawals for manufacturing (15 bgd), without development of any additional industrial water supply. Reuse by industry of an additional 14 percent of municipal and manufacturing effluents in 1980 would release enough potable water from manufacturing use to meet the 1965–1980 growth in municipal water withdrawal needs as well.

(b) By 2020, industrial reuse of 54 percent of total municipal and manufacturing effluents would meet the entire projected increase in industrial water withdrawals.

(c) By 2020, projected growth in municipal withdrawals could be met by a reuse of an additional 33 percent of municipal and manufacturing effluents. Adding this to the 54 percent referred to in (b) above would involve a total reuse of 87 percent of municipal and manufacturing effluents and obviate the need for any additional water supply. To accomplish this, some of the growth in municipal use would have to be supplied by reuse of municipal wastewater and, therefore, would depend upon solution of possible public health problems previously mentioned.[35]

One of the chief ingredients in this enhanced competitive position is the escalating cost of municipal water supply. The greater distances for high quality sources produce higher transmission and possibly litigation expenses. Closer lower-quality sources produce higher storage and water treatment costs. Storage costs are increasing because of the increasing cost and decreasing storage potential of available sites and the increased opportunity costs (costs absorbed because of the inability to realize more efficient uses) of flooding the land, while transmission costs are rising because of the increased distance from the source of supply and the high energy inputs that may be required.[36] The levels of component costs vary from city to city, of course, depending upon the particular mix of need. But the costs are rising rapidly everywhere.

The cost of reclaiming wastewater, by contrast, is declining relatively. The 1972 amendments to the Clean Water Act call for a massive effort to upgrade the extent and degree of wastewater treatment by municipalities and industries. As the quality of effluent from cities improves, the marginal cost required to convert the effluent into a useful and marketable product and to deliver it to the user (treatment plus distribution costs)—the true cost of water re-use—is declining. As cities project the costs of various supply alternatives, many are finding as a result of new pollution abatement requirements that at some point in the future water re-use and the traditional supply-augmentation cost curves are intersecting. In Denver, for example, estimates of future water supply costs have led the city to launch an ambitious water re-use research study, even though current comparisons reveal a cost advantage for traditional water sources.[37] Similarly, the Contra Costa County Water District near San Francisco plans extensive future re-use of municipal wastewater for industrial purposes, even though the current cost of potable water is less than for the treated effluent.[38]

The cost of reclaimed wastewater to the user will include the extra costs for treatment, conveyance, and storage. The intended use of the reclaimed water and the scale of operation will strongly affect treatment costs. Table 1.5 illustrates the costs of reclaiming wastewater to progressively higher levels for a one-, 10- and 100-mgd plant. The first two stages of advanced waste treatment drop from 58 to 15.6 cents per 1000 gallons as one moves from a one-mgd to a 100-mgd plant. Besides these obvious economies of scale, the 1972 amendments to the Clean Water Act, which eventually require nitrogen and phosphorus removal, will mean that the cost increment needed to make the effluent available for re-use will drop substantially. Finally, it should be pointed out that treatment costs are site-specific, strongly influenced by the particular character of the effluent, treatment-plant characteristics, and level of treatment needed. A comparison of treatment costs for existing water re-use projects (Table 1.6) reveals variation ranging from $2.77 per 1,000 gallons at Grand Canyon Village, Arizona, to $.02 at Florence State Prison, Arizona. The data suggest that even at present, reclamation costs ranging from 18 to 41 cents per 1000 gallons are not unusual.

The specific design of the wastewater reclamation and re-use system also provides ample latitude for attaining economic feasibility. With a wide range of prospective users, the water reclamation program could tailor the wastewater treatment to the specific customer need and make it available at varying prices. Large-volume users, such as industry or irrigators, could obtain water at lower costs while the entire system would benefit from delayed investment in new expensive water sources. Alternatively, if and when the health issues in wastewater renovation are solved, water re-use could function as a stand-by system to meet occasional peaking demands, thereby spacing out capital investment over more time.[39] Further, it is possible to devise methodology that internalizes social costs into models that test the feasibility of water re-use.[40]

37. R. D. Heaton, et al., "Progress toward Successive Water Use in Denver," Paper presented at the Water Pollution Control Federation Annual Meeting, Denver, 1974 (mimeographed).

38. Ralph Stone and Co., Inc., *Wastewater Reclamation: Socio-Economics, Technology, and Public Acceptance* (Springfield, Virginia, National Technical Information Service, 1974), p. VI–2.

39. For an example of this approach to water re-use, see Daniel Dworkin and Duane D. Baumann, *An Evaluation of Water Reuse for Municipal Supply* (Alexandria, Va., Institute for Water Resources, U.S. Army Corps of Engineers, 1974).

40. Stone and Co., pp. VI—17–22.

Roger E. Kasperson

TABLE 1.5

*Approximate Costs of Secondary and Advanced Treatment*[1]

| (June 1967 Cost Levels) | | | Costs of Advanced Treatment Processes in Addition to Costs of Secondary Treatment | | | | | |
|---|---|---|---|---|---|---|---|---|
| Capacity of Plant (m.g.d.) | Secondary Treatment[3] | | Nutrient removal (including suspended solids)[a 2 3] | | Removal of nutrients plus nonbiodegradable organics[2] | | Removal of nutrients & nonbiodegradable organics plus demineralization[b 2] | |
| | Capital Costs ($Million) | Total Unit Treatment Costs[c] (¢/1000 gal) | Capital Costs ($Million) | Total Unit Treatment Costs (¢/1000 gal) | Capital Costs ($Million) | Total Unit Treatment Costs[c] (¢/1000 gal) | Capital Costs ($Million) | Total Unit Treatment Costs[c] (¢/1000 gal) |
| 1 | 0.54 | 19 | 0.43 | 26.8 | 0.81 | 58 | — | — |
| 10 | 3.2 | 11 | 1.8 | 14.0 | 3.4 | 24 | 6.8 | 36 |
| 100 | 20 | 6.5 | 10.9 | 8.6 | 26 | 15.6 | — | — |

[a]Costs based on air stripping. If biological nitrification-denitrification is required, as is presently indicated, the costs would undoubtedly be greater. Costs of this process are not currently available, but some researchers have expressed the view that its use could raise the total cost of nutrient removal by as much as 40 percent.

[b]Based on assumed mineral concentration of 850 p.p.m. in effluent, reduced to 500 p.p.m. (drinking water standard), thus providing for one cycle of reuse. Costs of brine disposal, which may be substantial, are not included in above demineralization costs because of variability between sites.

[c]Includes operation and maintenance and interest and amortization on capital investment (at 4.5 percent interest over 25 years for comparative purposes only; not intended as a recommendation for financing assumption).

Sources of Cost Data:

[1]Jerome Gavis, *Wastewater Reuse,* prepared for the National Water Commission, PB-201-535 (National Technical Information Service, Springfield, Va., 1971).

[2]Robert Smith and Walter F. McMichael, *Cost and Performance Estimates for Tertiary Wastewater Treating Processes,* prepared for the Federal Water Pollution Control Administration, Report TWRC-9, Robert A. Taft Water Research Center (Cincinnati, Ohio, 1969).

[3]R. Smith, *Cost of Conventional and Advanced Treatment of Wastewater,* Journal of the Water Pollution Control Federation, 40 (1968), 1546–74.

Source: National Water Commission, *Water Policies for the Future*, p. 310.

41. See Okun, "Planning for Water Reuse," as well as his earlier article, "Alternatives in Water Supply," *Journal of the American Water Works Association*, 61 (1969), 215–224.

A central problem in wastewater reclamation is to discover ways of achieving wider re-use while, given the risks already discussed, ensuring the public safety in an economically and aesthetically feasible system. Or, to put it a different way, what should be the innovation process, given the risks inherent in the re-use of polluted waters, the stage of present scientific understanding, the relative lack of experience with planned re-use for high-order uses, and the differences among cities in wastewater produced and water supply options?

One approach that has influenced the thinking in this book is the notion of a "plural" water supply comprising a hierarchy of water quality. As advocated by Daniel Okun, this concept begins with the argument that barring a surplus, no higher quality water should be used for purposes that can tolerate a lower grade.[41] Instead of a single grade of water for all uses, a water supply could include several different grades in quality tailored to particular needs. Since potable uses constitute a small portion (perhaps 5–15 percent) of the total use in a city, managers could extend the water supply over a long period of time by reclaiming wastewater and recycling it for the lower-order uses that make up the bulk of demand. Such a system would produce water more cheaply than one employing a single higher grade and also would eliminate the health issues involved in direct, potable re-use. Meanwhile, increased experience with water reclamation and re-use would pave the way

TABLE 1.6

*Comparison of Wastewater Reclamation Costs at Selected Sites*

| Site | Volume | Type of Re-use | Treatment costs (per 1000 gallons) |
|------|--------|----------------|------------------------------------|
| Amarillo, Texas | 4.5 MGD | Factory cooling | $. 82 |
| Camarillo, California | 2.3 | Irrigation | .19 |
| Colorado Springs, Colorado | 21.0 | Irrigation; cooling | .35 |
| Denton, Texas | 1.5 | Cooling | .18 |
| Florence State Prison, Arizona | 0.68 | Crop Irrigation | .02 |
| Fort Carson, Colorado | 0.3 | Golf-Course Irrigation | .39 |
| Grand Canyon, Arizona | 0.22 (summer) | Toilet Flushing; Irrigation | 2.77 |
| Irvine Ranch, California | 2.8 | Pasture Irrigation | .41 |
| Lubbock, Texas | 2.8 | Cooling | .30 |
| Odessa, Texas | 5.5 | Cooling; Make-up Water | .67 |
| Prescott, Arizona | 1.5 | Golf-Course Irrigation | .09 |
| S. Lake Tahoe, California | 7.5 | Recreational Lake | .39 |

Source: Ralph Stone and Co., Inc., *Wastewater Reclamation: Socio-Economics, Technology, and Public Acceptance* (Springfield, Va., National Technical Information Service, 1974), pp. VI–7 to VI–8, appendix VII.

for a sound process of innovation. In fact, such systems have existed for some time in water-short areas like the Virgin Islands, the Bahamas, and Hong Kong.

There are problems, of course—particularly for older cities with established systems. A plural water supply would entail, for example, the danger that nonpotable water might enter the potable lines. Yet an ever-increasing number of technological means are available for reducing the risk, and even if cross-connection did occur for a period of time, the resultant health hazard might well be less than that of long-term ingestion of potentially toxic substances in a single, lower grade of water.[42] The costs of producing a second supply system connecting each household might well, in the absence of federal support, be prohibitive in many cities. Yet substantial expenditures to separate storm and sanitary sources might well have been more profitable if applied to plural water supply systems. Moreover, a recent study by Paul Haney and Carl Hamann suggests that the costs of dual systems may be much less than generally realized.[43] Even in older cities, certain savings are possible from density economics. In new urban areas it has been estimated that acceptable dual systems could be provided at a cost about 20 percent higher than that for a single-supply system.[44]

Despite the problems, plural water-supply systems provide a wide range of options for re-using reclaimed water. The siting of industrial parks or other

42. Okun, "Planning for Water Reuse," p. 621.

43. Paul D. Haney and Carl L. Hamann, "Dual Water Systems," *Journal of the American Water Works Association*, 67 (1965), pp. 1073–1099.

44. Daniel A. Okun and F. McTunkin, "Feasibility of Dual Water Supply Systems," Paper presented to the 7th Annual American Water Resources Association Meeting (October 1971).

45. National Water
Commission (note 1,
above), p. 314.

high-volume water users adjacent to wastewater plants would allow for direct piping. Or reclaimed water could be piped to nearby agricultural regions, golf courses, or green belts. Such systems would permit a discriminating pricing policy, with much higher charges for higher grades of water, thereby relieving pressure upon the purer water sources. Different lines could operate at the optimal pressure for the particular service. Such changes would, of course, provoke a small revolution in water management.

*Direct, potable re-use.* Direct potable use of renovated wastewater would trigger a considerably greater revolution. The scientific and professional community engages in acrimonious debate over the public health issues of direct, potable re-use, particularly when the system is a relatively closed-loop one. The National Water Commission in its final report, endorsing the research program requested by the American Water Works Association, strikes a cautious tone on re-use for potable purposes:

> The Commission believes that direct reuse of water for industrial purposes and that indirect reuse for purposes of human consumption will increase. Where feasible, such indirect reuse should be minimized by limiting wastewater reuse to processes that do not involve human consumption. This will have the effect of releasing for human consumption potable water now being used by industry. However, previously demonstrated successes in protection of public health in instances where municipal water supplies are derived from indirect reuse suggests that increases in such indirect reuse for human consumption should not be discouraged.

> In regions where a high-quality source of water is used for irrigation and cropped fields or recreation turf areas such as golf courses and a source of treated municipal wastewater is available, arrangements for water exchange should be considered. Nutrient-rich municipal wastewater would be used for irrigation and exchanged for high-quality water which could be used for domestic and industrial uses.

> Direct reuse of water for human consumption should be deferred until it is demonstrated that virological and other possible contamination does not present a significant health hazard. Further knowledge on the subject is necessary . . .[45]

The concern of the professional water and wastewater managerial community for direct, potable re-use is apparent in two policy statements released over the past several years by the relevant professional associations. In 1971 and then again in 1973 the American Water Works Association indicated that:

> The Association is of the opinion . . . that the current scientific knowledge and technology in the field of wastewater treatment are not sufficiently advanced to permit direct reuse of treated wastewaters as a source of public water supply, and it notes with concern current proposals to increase significantly both indirect and direct use of treated wastewaters for such purposes.

The Association believes that the use of reclaimed wastewater for public water supply purposes should be deferred until research and development demonstrate that such use will not be detrimental to the health of the public and will not affect adversely the wholesomeness and potability of water supplied for domestic use.[46]

In November of 1973 a joint statement of the American Water Works Association and Water Pollution Control Federation echoed this caution: "there is insufficient information about acute and long-term effects on man's health resulting from such uses of wastewaters . . . [and] fail-safe technology to assure the removal of all potentially harmful substances from wastewaters is not available."[47]

The Environmental Protection Agency, the institution that holds the key to any future direct potable re-use, enunciated its policy position in 1972:

We do not have the knowledge to support the direct inter-connection of wastewater reclamation plants into municipal water supplies at this time. The potable use of renovated wastewater blended with other acceptable supplies in reservoirs may be employed once research and demonstration has shown that all the following conditions would be met:
(a) Protection from hazards to health;
(b) offers higher quality than available conventional sources;
(c) results in less adverse ecological impact than conventional alternatives;
(d) is tested and supplied using completely dependable chemical and biological control technology;
(e) is more economical than conventional sources;
(f) is approved by cognizant public health authorities.[48]

This policy, if rigidly applied, could be restrictive for future potable re-use of renovated wastewater. The new drinking water standards eventually adopted by the EPA could do much to determine the course of water re-use in the United States. But the interim standards released in the spring of 1975 showed no change in this position, and it appears unlikely that the final standard will specifically consider re-use.

Whereas these policy statements concentrate on deficiencies in the current state of scientific knowledge and treatment and fail-safe technologies, other informed observers have also cited (1) inadequacies in operator training and performance and supervision by regulatory agencies, (2) a lack of investment in maintenance of plants, and (3) deficiencies in the technology and resources for analysis and routine monitoring of potable waters drawn from polluted sources.[49] These issues apply as well, it should be noted, to existing water supply systems. All sources agree on the need for an extensive national research and demonstration program on the safety issues connected with potable re-use. Current research at Denver, the Dallas Water Reclamation Center, and EPA laboratories and projects should provide important results.

46. American Water Works Association, "On the Use of Reclaimed Wastewater as a Public Water Supply Source, AWWA Policy Statement," *Journal of the American Water Works Association*, 63 (1971), 609; "Use of Reclaimed Wastewaters as a Public Supply Source," *Journal of the American Water Works Association*, 65, Pt. 2 (1973), 64.

47. American Water Works Association and Water Pollution Control Federation, "Joint AWWA-WPCF Statement," *Journal of the American Water Works Association*, 65 (1973), 700.

48. Schmidt and Clements, p. 94.

49. Okun, "Planning for Water Reuse," pp. 619–620.

**Roger E. Kasperson**

## Study Objectives

Four main questions pervade the dissemination of water re-use systems in American cities:

(1) Is the reclaimed water safe for the purpose for which it is intended?
(2) Is water re-use economically feasible?
(3) Will the public accept reclaimed water for its use?
(4) Will the managers recommend the adoption of a water re-use system?

The first question has received brief treatment earlier in this chapter. Safety is a critical consideration that merits a major national research effort; the present book explores the state-of-the-art of health hazards involved in direct potable re-use. It will be some years before definitive judgments may be made, but it is important all the same to delineate the major issues as well as the scale of risk in order to evaluate the near-term prospects for water re-use by American cities.

Despite the obvious importance of economic feasibility issues, there has been remarkably little research on this topic. Most of the work to date deals with preliminary cost data and projections from advanced waste-treatment pilot plants. Precious little research has focused on the complexity of calculating benefits or the feasibility of alternative system designs. In particular there is a need for more effective strategies to deal with seasonal and diurnal peaking as it relates to risk and the staging of capital investment.

The issue of public acceptance of renovated water for high-order uses first stimulated the research for this book. In light of the fluoridation conflagration, many managers and professional experts have feared that public opposition would emerge as a paramount obstacle to the adoption of re-use systems in American cities. Given the absence of experience, however, knowledge of the attitudes and behavior of the public on this issue is meager at best. There is, for example, a need to know what variables influence levels of acceptance. Does acceptance vary with the particular use of the envisioned processed effluent? Is it possible to identify underlying psychological determinants? What is the role of education and actual experience with reclaimed wastewater? If these issues can be resolved with any precision, it will be possible to tackle strategies for preparing the public for water re-use. An emerging body of studies analyzes this potential obstacle.

Finally, the behavior of the managers of the water supply and wastewater treatment systems is critical to the success of water re-use. The decision to adopt water re-use is for most cities a complex one involving a variety of actors from local, regional, and state agencies. In addition to worrisome health hazards, professional biases, past training and experience, and organizational goals all may play some role in managerial behavior. Although the technology connected with advanced waste treatment has, for the most part, existed for some time, the movement to water re-use as a means for solving water supply shortages surely represents a managerial revolution. The managers occupy a critical link in the decision process—if they fail to introduce and/or support the innovation, adoption is unlikely. An analysis of their attitudes and likely behavior is, therefore, a central objective of this study.

Such questions pose interesting methodological and research difficulties. Customarily, studies of the diffusion of innovations tap a record of adoptions available for behavioral analysis. In the case of water reclamation and re-use there is a history of adoption basically only for the lower-order and therefore noncontroversial uses. For the higher-order uses the innovation process is still in the pioneering stage. The problem is not unlike that plaguing recent studies of "nondecision-making"—namely, how to analyze the "non-event" in American cities.[50] Inevitably in the re-use situation, decision-making and attitude formation, the results from preliminary experimentation or demonstration projects, analogue situations, and inevitably an assessment of, given time, what the response of managers would be, all must provide the data for inferences. Obviously, this is less than ideal, from the research perspective, but necessary given the stage of water re-use innovation. One works with what is available.

Research began in 1969 with the aid of a grant from the Office of Water Resources Research (currently the Office of Water Resources and Technology). To assemble comparative data on the obstacles to community adoption of water re-use systems, research teams visited eight cities in the United States during the summers of 1970 and 1971. The eight cities were Gloucester, Massachusetts; Wilmington, Delaware; Kokomo and Indianapolis, Indiana; Lubbock and San Angelo, Texas; Santee, California; and Colorado Springs, Colorado. At each study site, researchers gathered detailed information on water supply and wastewater treatment, managerial attitudes, and public acceptance. Eventually they conducted some 625 interviews by random cluster sample from the general public and 45 more detailed interviews with public officials and professional managers. Later chapters will provide more detailed information on sample and instruments. Briefer reconnaissance studies conducted at Lake Tahoe, Los Angeles, and Lancaster, California, and Tucson, Arizona, supplemented the more detailed studies at the eight study sites.

The analysis of economic feasibility benefits from one of the most developed water re-use systems in the world—the ambitious and imaginative wastewater reclamation program in Israel. Since a colleague (Stephen Feldman) has conducted primary research on this system, his findings mesh nicely with studies in the United States and suggest implications for evaluating re-use feasibility within the framework of the water economy of Israel (Chapter 6).

The analysis of managerial behavior involved the administering of lengthy interviews and psychological tests among public health officials and consulting engineers. The requirements of this sample differed from those aimed at managers at the city level. Both region and the type of consulting firms guided the selection of the sample, which eventually embraced some 98 consulting engineers and 22 public health officials from some 33 engineering firms and 10 state public health departments. The results provide the base for the discussion of managerial behavior.

50. Peter Bachrach and Morton S. Baratz, *Power and Poverty: Theory and Practice* (London, Oxford University Press, 1970). Matthew Crenson, *The Un-Politics of Air Pollution* (Baltimore, Johns Hopkins Press, 1971).

The chapters that follow will provide a detailed discussion of methodology, including sample and research instruments. Figure 1.2 constitutes an overview of the empirical data collection efforts.

## Organization of the Volume

The book comprises three parts, in which the bulk of the discussion focuses on the four major potential obstacles to the growth of water re-use for supplying new water for American cities. In Chapter 2 Roger Kasperson gives a concise historical overview of the development of water re-use and devotes particular attention to the United States experience available for future re-use innovations. The chapter concludes with an examination of the current role of the Federal government. In Chapter 3 David McCauley describes the experience with water re-use in the six American cities which served as major study sites in this project.

Part II defines the potential obstacles to the diffusion of water re-use systems. Reviewing the literature and range of studies on this subject, John Reynolds in Chapter 4 identifies the major public health problems and the types of risk each involves. He then examines the implications of these findings for water re-use development in the United States.

The next two chapters are concerned with issues of economic feasibility. Daniel Dworkin (Chapter 5) proposes a simulation model to evaluate the economic competitiveness of water re-use. In Chapter 6 Stephen Feldman, drawing upon previously unpublished data, examines the development of

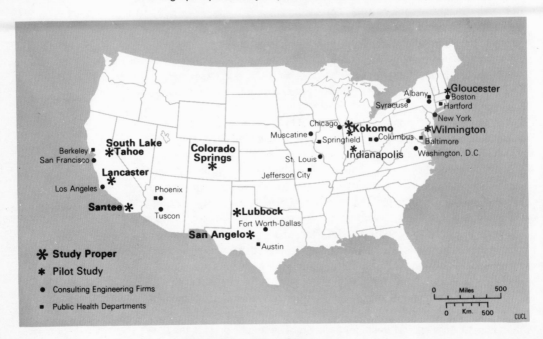

Figure 1.2: Distribution of Study Sites.

water re-use schemes in Israel. Evaluating the economic feasibility and public acceptance of water reclamation and re-use within the water economy, he suggests a number of the economic and institutional considerations which have contributed to the success of the re-use program in Israel.

Chapter 7 centers upon the public acceptance problem. Drawing upon the results of other studies, the Clark University study of attitudes toward reclaimed water, and the limited actual experimentation with water re-use both in the United States and abroad, Kasperson analyzes at length the bases for attitudes toward renovated wastewater and the role that education and experience may be expected to play. He concludes that urban Americans are basically rational in their attitudes toward water re-use, and that there are grounds for optimism toward cautious deployment of water reclamation and re-use technology over the long run.

Next is an analysis of managerial behavior as it relates to water re-use decisions. In Chapter 8 Thomas Nieman does for consulting engineers and public health officials what Chapter 7 does for the general public—he delineates their attitudes toward the adoption of renovated wastewater for potable supply and then analyzes the social-psychological bases of these attitudes.

Part III summarizes what the findings of the various chapters of Part II suggest for the process of innovation and public policy. In Chapter 9 David McCauley, drawing upon theories of adaptive behavior, formulates a general model of managerial adaption to the water re-use innovation. Focusing on environments, goals, and strategies of managerial systems, he proposes a three-stage model of the decision process. Finally, Chapter 10 summarizes the major findings of the study and makes policy recommendations.

# The Past as Prologue: Water Re-Use in the United States to 1975

Roger E. Kasperson

**2**

For the modern American city, water re-use is a significant, perhaps even radical, departure from the time-tested means of augmenting water supply. Debates rage at professional meetings over issues of safety, economic feasibility, and public response. Recognized experts quarrel over the relative merits of land versus water disposal, the problems of carcinogens in water supply, alternative uses of reclaimed wastewater, the relevance of the Santee experience, the impact of the Clean Water Act Amendments, the practicability of dual or direct piping of reclaimed water. And the incumbents of municipal water supply and sewage disposal departments, together with their colleagues in state agencies and consulting firms, fret about the nagging uncertainty inherent in new technologies and new institutions. Everywhere rapid change threatens the professionalism bred by day-to-day expertise. At first glance this skittishness is puzzling; it seems out of proportion to the adaptation involved, for the concept of water re-use is no stranger but rather a familiar, if not always welcome, guest. After all, there is little new in the notion of re-use.

Prior to the rise of cities, some nomadic and pastoral societies routinely returned simple wastes to the earth, where they not only failed to disturb the ecosystem but, indeed, replenished local flora and fauna. The same might be said for the network of septic tanks in more contemporary rural societies. It is only with the rise of the modern city that the complexity and volume of wastes have reached such formidable proportions. At first the city dweller tried to live with his wastes, storing them until conditions became intolerable.[1] Then he was forced to adapt—to expend energies to haul the waste away and distribute it on the land. It soon occurred to some enterprising official or entrepreneur that this unpleasant situation might have some unforeseen benefits. Why not use this waste as fertilizer and even as irrigation water?

A waste disposal–irrigation system may have operated from Bunslau, Prussia, beginning in 1559 and continuing for the next 300 years.[2] About the same time, German farmers were utilizing municipal wastes from Berlin on surrounding farmlands. In Scotland, Edinburgh distributed its sewage to nearby agricultural areas for spreading as fertilizer. Berlin purchased lands for sewage irrigation in 1869. By the middle of the nineteenth century, it was clear that cities would need a relatively quick and efficient method to dispose of waste if they were to avoid the accumulation of waste and epidemics of diseases. This fact led the Royal Commission on Sewage Disposal (England) to conclude in 1857 that "the right way to dispose of town sewage is to

1. Richard E. Thomas, "Land Disposal II: An Overview of Treatment Methods," *Journal of the Water Pollution Control Federation*, 65 (1973), 1477.

2. A. M. Buswell, et al., *The Depth of Sewage Filters and the Degree of Purification*, "Illinois, Division of State Water Survey Bulletins," 26 (Springfield, Illinois, The Author, 1928).

3. Quoted in Thomas.

4. Richard H. Sullivan, Morris M. Cohn, and Samuel S. Baxter, *Survey of Facilities Using Land Application of Wastewater* (Washington, D.C., U.S. Environmental Protection Agency, 1973), p. 170.

5. Leonard B. Dworsky, *The Nation and Its Water Resources*, Part I (Washington, D.C., National Technical Information Service, 1962), p. 40.

6. For a discussion of water-borne epidemics during this period, see Nelson M. Blake, *Water for the Cities* (Syracuse, Syracuse University Press, 1956), pp. 248–264.

7. Duane R. Egelund, "Land Disposal I: A Giant Step Backward," *Journal of the Water Pollution Control Federation*, 65 (1973), 1466.

apply it continuously to land and it is only by such application that the pollution of rivers can be avoided."[3] The first solution was simply to collect the waste in these primitive combined sewer systems and to dispose of them in nearby rivers—which immediately, of course, became septic. So the land-disposal means of controlling waste, sometimes accompanied by planned irrigation and fertilization of agricultural land, became the accepted wisdom during the latter part of the nineteenth century.

In the United States, the large eastern cities did not join this movement, but since the turn of the century, cities in California, Montana, Utah, and Wyoming have shipped their sewage to surrounding farmlands where irrigation, ground-water supplementation, and nutrient enrichment were all tangible benefits. Woodland, California, for example, initiated sewage farming for hay and pasture land in 1889. San Antonio, Texas, began to irrigate 3,000 acres of land with sewage effluent in 1900.[4] Meanwhile, Hanford and Fresno, California, established similar systems employing raw sewage. During the latter part of the nineteenth and early twentieth century, many other small land-disposal and re-use systems sprang up in a number of cities in the West and Southwest.

But changes were already under way which quickly overtook this approach to waste management. Between 1860 and 1900 the United States experienced a rapid growth of urbanization (Table 2.1). Whereas in 1860, 20 percent of the nation was urban, by 1900 this percentage had doubled. Municipal water-supply systems grew at an even faster pace, increasing from only 136 in 1860 to over 3,000 by 1900.[5] To meet the burgeoning accumulation of waste, cities chose the most convenient disposal outlet—the nearby rivers. Frequently this discharge of sewage was close enough to water intakes to permit pathogens to enter the water-supply source. This deterioration of sanitary conditions, coupled with inadequate water treatment, contributed to water-borne epidemics in nineteenth-century American cities.[6] In 1880 the death rate from typhoid fever, for example, was 31.9 per 100,000 population for New York, 57.6 for Philadelphia, 42.4 for Boston, and 59.0 for Baltimore. In Chicago, which drew its water from Lake Michigan, there were 2,000 deaths (a death rate of over 173 per 100,000) from typhoid fever alone. Along the polluted Merrimack River in Massachusetts, typhoid death rates in Lowell and Lawrence soared to over 100 in 1890. Cholera has a similar history.

With this linkage of water pollution to public health, the science of water pollution control took its first important strides. Despite the wide acceptance of filtration in Europe, American cities eschewed it until the latter part of the nineteenth century. In 1874 Poughkeepsie, New York, installed the first slow sand filter. The rapid sand filter followed between 1880 and 1885. By 1875 waste-treatment processes such as sedimentation and chemical precipitation were beginning to show satisfactory results. Philadelphia (about 1910) was one of the first cities to use liquid chlorine for disinfection. The period, in short, saw the foundation of many of the contemporary treatment processes—septic tanks, Imhoff tanks, contact beds, trickling filters, and activated sludge.[7] The institutional structure also emerged. In 1869 Mas-

TABLE 2.1

*Size Distribution of Urban Places, 1860–1900*

| Year | Urban Places Total | Over 100,000 | 50,000– 100,000 | 25,000– 50,000 | 10,000– 25,000 | 2,500– 10,000 |
|------|------|------|------|------|------|------|
| 1860 | 392 | 9 | 7 | 19 | 58 | 299 |
| 1900 | 1,737 | 38 | 40 | 82 | 280 | 1,297 |

Source: U.S. Department of Commerce, Bureau of the Census, *Historical Statistics, 1789–1945*, Series B 13–23, p. 25.

sachusetts established the first state board of health, and the other states were quick to follow.

These developments slowed the movement to land disposal of wastewater. Instead, treatment facilities with subsequent water disposal proliferated throughout most of the nation. While land-disposal systems continued to prove their effectiveness and to grow, particularly in the West and Southwest, they were not to re-emerge as a major alternative for large industrial cities until the 1970's, when the original focus upon an economical means of waste disposal would shift to something quite different. Meanwhile, however, significant water re-use projects began to appear in a number of American cities.

# Planned Re-Use for Nonpotable Purposes

Reclaimed water can be used for crop, forest, and golf-course irrigation, for industrial purposes, for recreation, and finally for potable uses. Any discussion of wastewater re-use innovations during the latter part of the twentieth century must, of course, take into account the extant patterns of re-use, the record of accomplishment to date, and the nature of the problems experienced. The most significant developments have, to date, involved the nonpotable uses of reclaimed water.

*Irrigation through land disposal.* In 1971 there were 571 municipalities supplying effluent for irrigation uses, mostly in areas of low annual rainfall. Generally, public health officials consider sewage with primary treatment safe for irrigating cotton, beets, or vegetables grown for seed production, for animal feed and pasture crops, and for woodlands. Properly disinfected, treated sewage which meets bacterial standards equivalent to drinking water standards is appropriate for all crops. Virtually no cities currently discharge raw wastewater onto land areas.

The application of wastewater to land areas may utilize a variety of methods—spray irrigation, flooding, ridge and furrow, and subsurface irrigation. Of these, spray irrigation appears to be the most common. It is impor-

8. See, for example, R. R. Parizek, et al., *Waste Water Renovation and Conservation*, Penn. State Studies, 23 (University Park, Pennsylvania State University, 1967); Stanley P. Pennypacker, William E. Sopper, and Louis T. Kardos, "Renovation of Wastewater Effluent by Irrigation of Forest Land," *Journal of the Water Pollution Control Federation*, 39 (1967), 285–296; Louis T. Kardos, William E. Sopper, and Earl A. Myers, "A Living Filter for Sewage," *Yearbook of Agriculture, 1968*, ed. U.S. Dept. of Agriculture (Washington, D.C., 1969). pp. 197–201.

9. Kardos, et al., p. 201.

10. Thomas, p. 1480.

11. The Environmental Protection Agency now publishes these data. See, for example, W. A. Hutchins, *Municipal Waste Facilities in the United States* (Washington, D.C., Environmental Protection Agency, 1972).

12. Thomas, p. 1477.

13. U.S. Department of Health, Education, and Welfare, Public Health Service, *Summary Report: The Advanced Waste Treatment Program* (Washington, D.C., 1965), p. 118.

14. Sullivan, et al.

tant to note, of course, that the most common approach to management of waste disposal outside of cities and in developing urban areas employs septic tanks—soil absorption systems which serve some 14 million homes, or 24 percent of the total population, in the United States and discharge up to several bgd into the soil.

The most important single project involving land application and re-use has been conducted at Pennsylvania State University.[8] In 1962 a group of scientists and engineers inaugurated an effort to determine whether the university's treatment plant could spray its wastewater onto the land to support crops and forest land instead of into a small mountain stream, where it was causing water quality problems. Researchers have carefully monitored and regularly reported on this study since 1962. The results have been encouraging for water reclamation through land disposal. Spreading over forest lands decreased the alkyl benzene sulfonate (ABS) found in hard detergents below United States Public Health standards for drinking water, effectively removed phosphorus and, with greater soil depth, removed equal amounts of nitrate nitrogen, potassium, calcium, magnesium, and sodium. The researchers conclude that only 129 acres of land would be needed to dispose of and reclaim one mgd of wastewater, approximately the output of a community of 10,000 persons.[9] This work has provoked much interest in the use of soil and its microbes and vegetation cover as a "living filter" for wastewater reclamation.

Melbourne, Australia, provides another important model for modern large-scale land disposal irrigation systems.[10] In operation since the end of the nineteenth century, this 115-mgd water re-use system involves crop irrigation during the summer season, spray-runoff in the winter season, and treatment lagoons during the wet season. The irrigation supports an 11,000-hectare livestock farm. The low operating costs combined with a 95 percent biochemical-oxygen-demand (BOD) removal and 92 percent suspended solids provide ample reason for considering land-application systems for middle-size and large cities if nearby inexpensive land can be found.

Since 1940 periodic inventories of municipal waste facilities have furnished information concerning cities applying wastewater to the land.[11] Table 2.2 shows the trend between 1940 and 1972. The data indicate both the continuing spread of land-disposal systems and the increase in recent years in system size. Such systems on the whole are, however, still small and serve an average population of only 11,000. Furthermore, 316 of the 571 systems in 1972 involved crop irrigation in the 13 western states, whereas the remaining 255 systems involved infiltration or unidentified methods of application in the other 32 states.[12] Another source notes that more than one fourth of the 800 municipal plants in Texas provide effluents to irrigate cotton or cattle feed crops.[13]

A survey by the Environmental Protection Agency in 1972 involved on-site investigations of 67 community and 20 industrial land-disposal systems, mail surveys of 86 community and 35 industrial land-application systems, state health and water-pollution control regulations governing these projects, and surveys of foreign experiences.[14] A number of significant findings emerge.

TABLE 2.2

*Municipalities Using Land Application of Wastewater*

| Year | Number of Systems | Population Served (in millions) |
|------|-------------------|--------------------------------|
| 1940 | 304 | 0.9 |
| 1945 | 422 | 1.3 |
| 1957 | 461 | 2.0 |
| 1962 | 401 | 2.7 |
| 1968 | 512 | 4.2 |
| 1972 | 571 | 6.6 |

Source: Richard E. Thomas, "Land Disposal II: An Overview of Treatment Methods," *Journal of the Water Pollution Control Federation*, 65 (1973), 1477.

15. Curtis J. Schmidt and Ernest V. Clements, III, *Demonstrated Technology and Research Needs for Reuse of Municipal Wastewater* (Cincinnati, Ohio, Environmental Protection Agency, 1975), p. 121.

Since land-application systems tend to locate on marginal lands of low value, relative isolation and the availability of necessary acreage are key factors in site selection. Consistent with the findings of other studies, the results revealed that land-disposal systems tend to be very small. Fully 73 percent of the systems surveyed, in fact, had capacities under five mgd and disposal sites of under 200 acres. The irrigated land ranged from no ground cover to cultivated crops and forest land, but grass predominated. Finally, the adequacy of environmental monitoring and regulation does little to inspire confidence. Most states lacked any set policies for handling the health and environmental impacts, failed to impose specific conditions, and seldom inspected the systems or required reports. Despite the lower-order use involved, it is exactly this flaw in surveillance and monitoring which troubles many professionals in water and waste management.

Another study of 132 land-disposal systems by Soil Conservation Service (SCS) engineers concluded that most municipalities look upon irrigation operations as primarily a means of effluent disposal; in fact, only 25 percent even charged a fee for the water supplied. Revenues from the re-use were estimated as comprising less than one percent of the treatment costs incurred by the municipality.[15]

Finally, although crop and forest irrigation certainly constitutes the bulk of the land application of effluent and reclaimed water, there have been some interesting cases of golf-course, park, and green-belt irrigation. Since 1924, for example, the Grand Canyon National Park has utilized municipal wastewater from an activated sludge plant for lawn sprinkling. Golden Gate Park in San Francisco has since the 1970's used a mixture of well water and sewage (reclaimed since 1932) to irrigate about 800 acres of park land. Among cities which have used or presently utilize wastewater for golf courses or parks are El Cajon and Palo Alto, California; Marine bases at El Toro and Pendleton, California; Los Alamos, Santa Fe, Gallup, Carlsbad, and Jal, New Mexico; Winslow and Prescott, Arizona; Sunnyside, Utah; Disneyworld in Florida; and a variety of luxury hotels and real estate developments in Las Vegas, areas of California, and the Virgin Islands. Perhaps the most important of

16. Ibid.
17. Ibid. pp. 224–229.
18. Ibid. pp. 246–253.
19. Ibid. pp. 266–273.

these, however, are Colorado Springs, Colorado (discussed in detail in Chapter 3), and St. Petersburg, Florida, where the use of reclaimed water to irrigate such facilities as golf courses, parks, lawns, and cemeteries serves specifically to reduce the demand on the potable water supply. These are prototypes of the "plural" water supply discussed in the preceding chapter.

*Re-use for industry.* The Survey of the SCS engineers identified only 15 industrial plants in the United States re-using municipal wastewater.[16] Although not nearly so widespread as land disposal for irrigation, the use of reclaimed municipal sewage for industrial purposes—for example, cooling, boiler feed, and copper mining, with cooling the clearly predominant use— dates back many years. Projections indicate that industry and steam electric power require an increasing proportion of the total water supply of the United States (Table 1.2), and re-use of both industrial and municipal wastewater is becoming a pressing need, particularly in water-short areas. Moreover, even partial implementation of PL 92-500 (the Federal Water Pollution Control Act Amendments of 1972) will make available and highly accessible to industrial plants significant new sources of high-quality water. Direct piping to high-volume users, one of the assumptions of a plural water supply system, should become economically very attractive.

Perhaps the best known example of industrial use of municipal wastewater is the Bethlehem Steel Company Sparrow Point Plant in Maryland. In 1942, after its deep well had become contaminated from salt water intrusion, the steel mill first contracted for 20 mgd of secondary treated effluent, which quickly proved superior to the saline Chesapeake Bay or available groundwater alternatives. Because of its once-through system, the company successfully uses a relatively poor quality effluent. The Baltimore plant supplies an estimated 120 mgd at the bargain price of one cent per 1,000 gallons. The Bethlehem Steel plant is now considering future improvements which include greater water re-use.[17]

Another well-known case is that of the Cosden Oil and Chemical Company at Big Spring, Texas. Beginning in 1943, Cosden contracted for the full rights for water from the Big Spring contact aeration plant. Currently, the chemical company uses some 0.5 mgd at a cost of about 34¢ per 1,000 gallons. Used as boiler feed, the water is of a quality superior to existing wells and satisfies approximately 25 percent of the company's total water need.[18]

Use of municipal effluents for industrial process water is largely concentrated in the Southwest and West. At Amarillo, Texas, a gasoline refinery and a power plant together purchase 4.5 mgd of municipal effluent; petroleum and chemical plants at Odessa, Texas, use 5.5 mgd of effluent for cooling water, paying in return virtually all the operating expenses of the municipal plant.[19] The Nevada Power Company uses Las Vegas wastewater for cooling water in its steam-electric power plant. Copper plants at Hurley and Santa Rita, New Mexico, use municipal effluent for process makeup water and copper recovery. In Burbank, California, a water reclamation plant has provided two mgd of very high quality cooling water for the city's steam electric plant since 1967. Moreover, the high degree of coordination between industry and the wastewater reclamation managers has been exemplary. At Sioux Falls, South Dakota, effluent supplies a nearby nuclear power plant.

But this record of re-use adoption lags seriously behind the potential. The Southwest has led the way because of high water costs and the poor quality of available water sources. But water in many American cities has been cheap and, given the pricing structure, particularly so for large volume users. This situation is changing, and in some industrial sectors, such as steam-electric power, future water needs will be extensive and the cost much higher. Clearly, industrial re-use presents one of the most attractive possibilities for water re-use in American cities. Burbank may serve as a model for future innovation of this type.

*Ground-water replenishment.* In many areas of the United States, bo⸍ and humid, ground-water levels are falling. Natural means for reple⸍ are not keeping pace with withdrawals. In California artificial re⸍ ready an important asset. Ground-water replenishment throᵤ⸍ represents an important option for arid areas of the West ⸍ the densely urbanized portions of humid regions.

Water re-use through ground-water recharge employ⸍ ent methods. Holding basins, in which a series of reservoiᵣ⸍ another and thereby impound the reclaimed water, constitute⸍ rangement which has influenced the ambitious re-use projects aᵤ⸍ Lubbock (see Chapter 3). A second approach, surface spreading, sᵣ⸍ the effluent over the ground at small depths, thereby avoiding disruption of vegetation cover and underlying soil, and allows it to percolate down to the water table. Frequently, specially designed and constructed spreading basins are employed, such as at Santee or Whittier Narrows, California. At Muskegon, Michigan, a land-disposal reclamation project utilizes spray irrigation as the spreading device.

Another important method utilizes injection wells to pump treated wastewater down into a depleted aquifer. This system is ideally suited to the recharge of deep, confined aquifers where surface spreading would be impractical, where the aquifer characteristics are especially favorable, and where land is prohibitively expensive.[20] The resulting reclaimed water can be used for augmenting general water supply or for preventing salt-water intrusion, as in Nassau County, Long Island, and Manhattan Beach, California. Injection wells do necessitate greater pretreatment for health reasons, however, because the wastewater does not enjoy the purification that percolation through the soil provides in surface-spreading schemes. At Orange County, for example, highly resistant soluble organic material has caused odor and taste problems in the 545-foot confined aquifer.[21]

Ground-water recharge utilizing surface spreading has important safety advantages over other forms of more direct cycle re-use as well as aquifer injection. In large industrial cities subject to variable periods of highly concentrated toxic flows, the toxic effluent can travel in "slugs" through the sewers and treatment plant. The chemical, biochemical, absorption, adsorption, and ion-exchange processes in the soil help to reclaim the effluent. Mixing in

20. George Tchobanoglous and Rolf Eliassen, "The Indirect Cycle of Water Reuse," *Water and Wastes Engineering*, 6 (1969), 36.

21. G. M. Wesner and D. C. Baier, "Injection of Reclaimed Wastewater into Confined Aquifers," *Journal of the American ⸍ Works Association,* ⸍ (1970), 210.

22. A. Amramy, "Re-Use of Municipal Waste Water," *Proceedings of the International Conference on Water for Peace*, 2 (1967), 423–424.

23. Warren Viessman, Jr., "Developments in Waste Water Re-Use," *Public Works, 96 (1965), 138*–200.

24. John D. Parkhurst, "Wastewater Reuse—A Supplemental Supply," *Journal of Sanitary Engineering*, ASCE, SA3, 96 (1970), 657.

25. John D. Parkhurst, "A Plan for Water Re-Use" (Los Angeles, 1963), p. 102 (mimeographed).

26. Private communication from Franklin D. Dryden, Head, Technical Services Department, County Sanitation Districts of Los Angeles, County.

the aquifer then further reduces any contaminants remaining. Finally, additional insurance is provided because the reclaimed water does not reach the withdrawal wells simultaneously.[22]

Economic feasibility depends upon a number of factors, including wastewater treatment processes, transmission to the recharge site, recharge facilities, and the costs of alternative sources. Increasingly, it is becoming clear that the transmission of wastewater and/or the acquisition of large tracts of accessible and technologically suitable land may be the major limiting factors. A study of these issues by Metcalf and Eddy, commissioned by the EPA, was, at the time of this study, nearing completion. The location and magnitude of the recharge site could well be the limiting sociopolitical factor.

The most significant adoption of re-use for ground-water replenishment is Whittier Narrows, in Los Angeles County, California. In 1962 the Los Angeles County Sanitation Districts established a 10-mgd reclamation plant which combined the activated sludge process and chlorination to produce a high quality wastewater for spreading and subsequent percolation along the San Gabriel and Rio Hondo rivers. After dilution with ground water, the renovated water returns from the aquifer and serves unrestricted uses, including municipal potable supply. Despite a slight increase in the mineral content of ground water from the recharge, careful monitoring and evaluation tests have satisfied the California State Water Quality Board that the renovated water has not contributed to public health risks.[23]

In 1970 John Parkhurst, Chief Engineer and General Manager of the Sanitation District, estimated that the operation of the Whittier Narrows plant over the first six and one-half years had returned to beneficial use some 95,811 acre-feet of good quality water, provided downstream savings totaling $1,575,787, and contributed some $728,445 to defray the original capital investment costs of the reclamation plant.[24]

The success of the Whittier Narrows plant contributed to the adoption in 1965 of a Los Angeles County Master Plan entitled *A Plan for Water Re-Use*,[25] which will provide wastewater treatment and re-use for the entire service area beyond the year 2005. The plan (Figure 2.1) involved the enlargement of the Whittier Narrows Plant (originally 10 mgd, but recently increased to 50 mgd) to an ultimate capacity of 125 mgd and the establishment of three additional plants—The Los Coyotes Water Renovation Plant with an initial capacity of 12 mgd, the San Jose Creek Plant with an initial capacity of 40 mgd, and a fourth plant at Long Beach. For the latter, officials in 1974 were evaluating the possibility of injecting reclaimed water into the ground through oil field repressurization wells, in order to improve oil recovery, park irrigation, and the prevention of sea-water intrusion through a ground-water barrier system.[26] In addition, the county districts operate a water reclamation plant that provides irrigation for landscapes and citrus groves at Pomona and a series of recreational lakes at Antelope Valley in Lancaster. (The next chapter discusses this latter project in depth.) Taken as a whole, the Los Angeles County efforts probably represent the most ambitious and important ongoing water re-use projects in the United States.

A noteworthy ground-water recharge program, similar in magnitude to Whittier Narrows, is presently under way in Tel Aviv, Israel. Since 1960

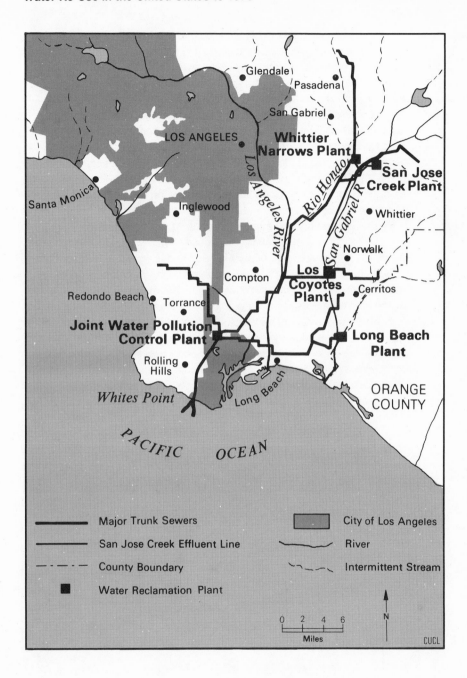

Figure 2.1: Water Reclamation and Re-Use in Los Angeles County.

27. John L. Rose, "Injection of Treated Waste Water into Aquifers," *Water and Wastes Engineering*, 5 (1968), 40.

studies have investigated the feasibility of reclaiming the large volume of effluent from the city instead of allowing it to empty into the Mediterranean. The existence adjacent to the city of large, vacant tracts of sand dunes underlain by impervious clay has permitted the development of a very economical wastewater reclamation system involving a series of anaerobic lagoons followed by secondary anaerobic, aerobic (two stages), and polishing lagoons. The effluent is then pumped to the dune area for ground-water recharge by surface-spreading. After mixing with existing ground water, the water enters a pipeline to mix with other sources prior to its conveyance to the south of Israel, where it will serve as general supply water. Eventually the reclamation project will involve some 35 mgd of Tel Aviv sewage. Chapter 6 discusses this case in considerable detail.

A more modest effort in a heavily urbanized, humid region occurred in Nassau County on Long Island. The population of Nassau County increased from 673,000 in 1950 to 1,500,000 in 1967 and is expected to reach 2,250,000 by the year 2010.[27] Since public water supply draws exclusively on ground-water sources, and since some southwestern sites had already lowered the water table sufficiently to cause salt-water intrusion, officials felt that steps were needed to prevent further encroachment. Consequently, the county constructed a .5-mgd demonstration tertiary treatment plant to provide renovated water, disinfected with chlorine and meeting United States Public Health standards for drinking water, for pumping to a nearby area which contained an injection well and other observation wells (Figure 2.2). The intent was to develop in Nassau County a hydraulic barrier across the western half of the southern edge of Long Island and, subsequently, to extend it to the eastern half. Initial operation of the demonstration plant was successful, and estimates placed the cost of renovated water from an enlarged 10-mgd plant at 14.7 cents per 1,000 gallons. There were continuing

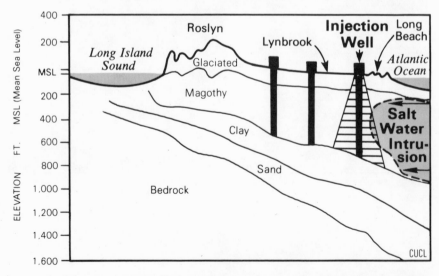

Figure 2.2: Ground Water Recharge at Nassau County. After John L. Rose, "Injection of Treated Waste Water into Aquifers," *Water and Wastes Engineering*, 5 (October 1968), 41.

problems with nitrogen in the reclaimed water, however, and EPA eventually withdrew its funding of the project. Recently a new County Commission won election on a campaign to halt ground-water contamination, and the project in 1975 was seeking support to shift to surface-spreading. Also, recovery from the drought of the early 1960's has convinced officials that the problem of ground-water depletion and salt-water intrusion is not as serious as originally thought.

Ground-water replenishment, like industrial re-use, offers a potential that far exceeds exploration to date. Perhaps the greatest advantage is the return to the indirect cycle method for subsequent natural purification and dilution prior to re-use. This spatial and temporal separation between treatment and use provides the important natural buffer to deal with the safety uncertainties discussed in detail in Chapter 4. Although there is some concern among public health officials that the movement of recharged water through the soil may be difficult to monitor and evaluate, preliminary experiences at both Whittier Narrows and the Pennsylvania State University project site indicate that given appropriate soil and ground-water conditions, this type of re-use promises vast potential.

*Re-use for recreation.* Though land disposal, industrial use, and even ground-water recharge of reclaimed water date back several decades or more, recreational lakes created from reclaimed sewage are of more recent vintage. Santee, California, is the pioneer, dating from the early 1960's, and has served as the model for more recent imitations at Antelope Valley (Lancaster), California, and Lubbock, Texas. (Since the following chapter describes each of these experiments at length, they will be noted only briefly here.) A more recent innovation at Michigan State University in East Lansing is also noteworthy.

Santee undertook its re-use program in 1959, creating in an arid region four artificial lakes fed by reclaimed wastewater from an activated sludge plant followed by natural infiltration through an old river bed and chlorination. An exceptional public education and publicity campaign allowed residents to participate gradually in recreation in a step-by-step progression—through lakeside picnics, boating, fishing where the catch could not be taken home, fishing with no restrictions on the catch, to swimming in a pool filled sometimes with reclaimed water and sometimes with other water resources. Recently, problems with solids and dissolved solids led the County Health Department and the Regional Water Quality Control Board to issue a cease-and-desist order. But during its life, the Santee experiment was the most visible and dramatic if not the most convincing water re-use project in the United States and has, as noted, spawned a number of successors.

At Lubbock, Texas, the "Canyon Lakes Project" created ornamental lakes and proposed two additional lakes to provide power boating and swimming. Yet even here, though recreation is unquestionably an important objective, the enhancement of land values is the primary goal. At the edge of the Mojave Desert, Antelope Valley, part of the Los Angeles County Water reclamation and re-use program, receives domestic wastewater with secondary and limited tertiary treatment from Lancaster for its new aquatic park. This park consists of three interconnected lakes, all designed to provide picnicking, camping, fishing, and boating.

In East Lansing, Michigan, a new 500-acre aquatic park with four lakes (with a surface area of 40 acres) currently utilizes .5 mgd of effluent from the East Lansing secondary treatment plant. Just beginning operation in 1975, the project calls for harvesting of aquatic plants, irrigation of nearby crops and forests, and, eventually, recreation on the lakes. A thorough exploration of public health issues, including an extensive monitoring program of bacteria and viruses, is now in process. Reclaimed water at the end of the lake chain serves for ground-water recharge and irrigation and may in the future upgrade the water quality in nearby streams.

These pioneering efforts with recreational uses of reclaimed water are important in several respects. First, they all combine a series of uses with exploitation of the natural purification processes of indirect cycle to reclaim wastewater. This is an important advantage over re-use with more direct cycles. Second, compared to earlier re-use efforts, these innovations involve a higher degree of sophisticated planning and monitoring to exploit varied uses of the water. They are more revolutionary, particularly in managerial terms, than they may appear. Third, they represent efforts which abbreviate or collapse, without creating undue hazards, the innovation timetable. Intensive scientific surveillance and evaluation begin to provide the experience necessary for the eventual potable uses of reclaimed water. They also demonstrate, and begin to exploit, the possibilities of a plural, or hierarchical, water supply.

## Models for Direct Potable Use

Chapter 1 suggests how volatile a subject the direct re-use of wastewater for potable supply is. Chapter 8 will provide insight into the intensity of feeling among water resources professionals, particularly public health officials, over this question. Crisis situations will increasingly tempt short-run emergency measures involving direct potable use, and more detailed considerations for long-run solutions will continue in water-short areas. But where precedents are lacking, innovations carry the price of great uncertainty. How does one substitute for experience? And where the innovations involve overriding questions of public health, where the reputation and competence of experts go on the line, taking risks does not come casually. Nor should it. However remote, however insignificant, the models of earlier times will affect significantly even (or perhaps especially) the pioneers in innovation.

*Chanute, Kansas.* From 1952 to 1957, Chanute, a town of 12,000 persons, experienced a series of severe water shortages, culminating in the summer of 1956 when the Neosho River, the town's source of water supply, ceased to flow. Officials considered various proposals but eventually decided in favor of direct re-use of the town's sewage. In a number of respects the situation was ideal—a relatively new secondary-treatment sewage plant, an effective water-treatment plant, a scant 5,000 feet of separation between sewage outfall and water intake, and a population accustomed to a poor quality municipal water. Figure 2.3 portrays the flow of treatment and re-use. Without public fanfare and without notification of the State Public Health Department, the town officials simply instituted the re-use scheme.

The system ran for some five months (October 14, 1956, to March 14, 1957) and recycled water through the town some eight to fifteen times. Although the water at all times met public health standards, it quickly deteriorated in quality: "The treated water had several objectionable characteristics. It had a pale yellow color and an unpleasant musty taste and odor. The tap water was high in chlorides, total solids, and organic content—adding to the taste and odor problem. Agitation of the water—as in drawing a glass of it from the tap—caused frothing because of the high ABS content."[28] In fact, had it not been for an ad hoc waste-stabilization pond that provided about seventeen days' average retention time (Figure 2.3), the quality of the water would have been considerably worse. In retrospect, it is doubtful that the re-use process could have continued more than a few weeks beyond the five-month period. Fortunately, rains put an end to the crisis and to the experiment.

The re-use period witnessed no known cases of water-borne disease or other adverse effects upon health. Special biological tests detected no pathogenic organisms or viruses in the water. Nevertheless, no systematic morbidity studies were conducted, and the situation in any event begs a number of health issues.

Public acceptance, though initially favorable, quickly deteriorated. Bottled-water sales flourished, citizens drilled some seventy new wells, others hauled water from neighboring towns and farm wells, and the city and the Santa Fe Railroad hauled and distributed gratis 150,000 gallons of drinking water. In short, it is unclear how many people actually drank the water. In retrospect, then, the Chanute adoption of water re-use succeeded only superficially. The measure did help in an emergency situation, but the limited technology and system design available in the mid-1950's did not serve to produce a water of satisfactory potable quality for the consumers, and Chanute is no model for present-day cities. It was an important event, nevertheless, calling attention as it did to the potential of direct potable re-use.

*Windhoek, Namibia*. Windhoek, located in an arid environment possessing few water-supply options, is the only city which (until recently) employed planned direct re-use for municipal potable supply. Consequently, it has considerable significance for the United States. In 1962 the city undertook research to develop advanced purification techniques to renovate the city's wastewater for direct municipal supply. After successful experience with a pilot plant, it instituted a one-mgd reclamation plant in 1968, and during favorable periods, then mixed the renovated effluent with the extant freshwater supply.[29] The water-renovation process employed conventional primary and secondary treatment followed by retention (to reduce bacteria, nutrients, and viruses), and several other processes, including algae flota-

28. Dwight F. Metzler, et al., "Emergency Use of Reclaimed Water for Potable Supply at Chanute, Kansas," *Journal of the American Water Works Association*, 50 (1958), 1031.

29. See G. J. Stander and J. W. Funke, "Conservation of Water Reuse in South Africa," *Chemical Engineering Progress Symposium Series*, 63, No. 78 (1967), 1–12; "Water Reclamation in Windhoek," *Scientiae*, 10 (1969), 3–14.

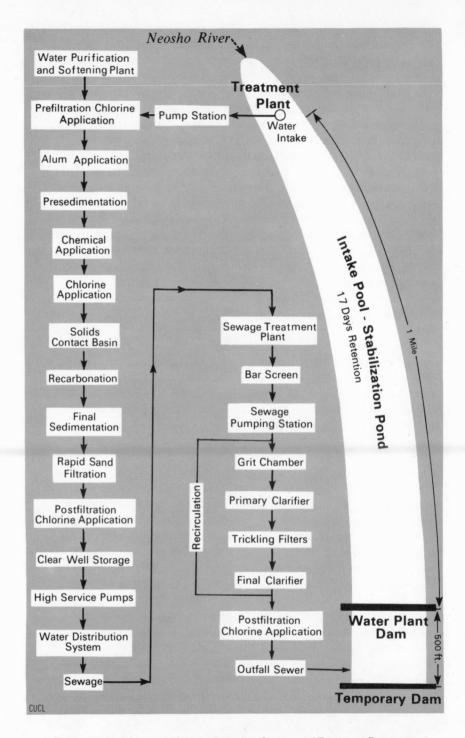

Figure 2.3: Flow Diagram of Water Recycling System and Treatment Processes at Chanute, 1956–57. After Dwight F. Metzler et al., "Emergency Use of Reclaimed Water for Potable Supply at Chanute, Kansas," *Journal of the American Water Works Association*, 50 (August 1958), 1021–57.

tion, free residual chlorination, rapid sand filtration, and activated carbon filtration. The actual degree of mixture with fresh water varied over time. When conditions were favorable, officials recycled effluent up to one third of the potable supply. At other times, when they could not meet chlorine demand, they did not recycle the wastewater.[30] During the first two years of its operation, the reclamation plant contributed 13 percent of the city's water consumption.[31] In 1971 plant operations were discontinued. In the spring of that year the quality of the maturation-pond effluent ceased to comply with water-quality specifications, and unusually heavy precipitation had filled water-supply reservoirs. Currently, officials are examining modifications and additions that will produce a higher grade of renovated wastewater.

It is important to note that several characteristics of Windhoek provided a very favorable context for this innovation. The wastewaters there contain no significant industrial wastes, so that chemical contamination is minimal. Second, the project involved a heavy investment in scientific and supervisory personnel which might not be available economically to later adopters. Finally, supply options were very few. Nevertheless, both the initial success and subsequent problems of the project suggest that direct potable re-use is still at the very experimental stage. Like Chanute, Windhoek demonstrates that it can be done. But it is also clear that a highly reliable, feasible system is still some time away.

*Denver, Colorado.* An important planning and pilot effort for eventual direct potable re-use is currently under way in Denver.[32] Population and water-use projections indicate a doubling of demand between 1970 and 2000, with water supplies, even with reapportionment of water from agricultural to municipal uses and import of transmountain waters, inadequate to meet these needs. With support from a major development-bond approval, Denver has established a plan calling for construction of a one-mgd tertiary plant to utilize secondary effluent from the metropolitan area. The goal will be "use increment removal" (water renovation as defined in the previous chapter), the restoration of the sewage as it leaves the water-treatment plants to its original quality. The city, in collaboration with University of Colorado scientists and State Public Health officials, will then launch an ambitious five to fifteen year study of the health effects. If the results satisfy city and state officials, the late 1990's will see a 100-mgd renovation plant with a nearly closed-loop system.

One of the interesting aspects of the Denver project, in addition to the extensive analysis of other reclamation plants, is the proposed use of a sophisticated computer model to analyze probable quality effects for variable blendings of recycled water with virgin waters. Annual, monthly, and daily bases of demand will determine the relative percentages of renovated water and virgin water that will be introduced into the municipal system.

Project sponsors envision the scheme as economically efficient. They estimate that the one-mgd pilot plant will produce water at $1.65 per 1000 gallons, while the 100-mgd plant would lower this figure to the 70–80 cents range. This would be highly competitive in price with other sources of supply using transmountain diversion.

30. Daniel A. Okun, "Planning for Water Reuse," *Journal of the American Water Works Association*, 65 (1973), 620.

31. Schmidt and Clements, p. 96.

32. See K. D. Linstedt, K. J. Miller, and E. R. Bennett, "Metropolitan Successive Use of Available Water," *Journal of the American Water Works Association*, 61 (1969), 610–615; R. D. Heaton, et al., "Progress toward Successive Water Use in Denver," paper presented at the Water Pollution Control Federation annual meeting, Denver, 1974 (mimeographed).

Less heralded than the Denver project but no less significant, the Dallas Water Reclamation Research Center has a similar purpose. Recognizing future water-supply needs of the city in 1995, this Center, with a one-mgd tertiary research plant, is undertaking studies designed to provide the basic scientific understanding which high-order uses of renovated wastewater will eventually require. The collaborative effort among the city, the U.S. Public Health Service, the Environmental Protection Agency Research Laboratory at Ada, Oklahoma, and other consultants represents the sound scientific work required now for future innovations. Phoenix, Arizona, is another city that might well be a candidate for early direct potable re-use.

This brief review of models for planned, direct potable re-use innovations indicates the limited accumulation of experience to date. Yet rudimentary as these attempts have been, they do suggest the feasibility of such systems for future urban water supply needs. Future innovations, however, will depend not only upon technological and scientific progress but upon the leadership of the federal government.

## The Role of the Federal Government

The federal government's involvement in water management, particularly as it bears upon water re-use, has been slow in coming. The significant legislation of the 1960's and 1970's has, despite its visibility, been evolutionary rather than revolutionary. To observers of domestic national policy it will come as no surprise that water policy has emerged in a piecemeal incremental fashion, in which interagency rivalries and conflicting priorities have shaped decisions as profoundly as have actual national needs. Moreover, the activity of different agencies promises to influence significantly the context within which American cities make decisions about water supply and waste disposal. Certainly federal leadership could well shape the type of re-use systems that emerge in American cities.

*Water supply*. Until very recently, the federal government has taken the stance that water supply is a matter for state and local government. The first significant developmental role emerged during the depression years when, casting about for useful projects, the Public Works Administration centered upon municipal water supply as one option. Eventually the program constructed some 500 water treatment plants costing $311 million and accounting for 50 percent of all new water-treatment plant construction between1933 and 1939.

The Water Supply Act of 1958 was the first major step, however, toward increased federal responsibility in municipal water supply. By providing greater utilization of water for local supply in federal multipurpose reservoirs and permitting federal agencies to plan and store water for anticipated future requirements of cities and industries, the legislation involved federal agencies in an unprecedented way in enlarging and improving the quality of urban water supply.

Several national assessments were also instrumental in calling attention to long-run issues in water supply provision. The Water Resources Committee of the National Resources Committee in 1935 and its successor, the Interagency River Basin Committee, utilized the river basin framework to formulate integrated water development patterns. The President's Water Resources Policy Commission in the early 1950's recommended an increased federal role in expanding municipal supply. In 1960 the Senate Select Committee on National Water Resources, and later the National Water Commission, provided comprehensive reviews of national water needs, calling attention in both cases to the critical importance of increased water re-use (see Chapter 1). The Water Resources Planning Act of 1965 provided continuing funds to support research and planning activities in water supply.

Perhaps the major role of the federal government over the years, however, has been the involvement of the U.S. Public Health Service with drinking water standards.[33] In 1914 the Service issued the first national water standards, which were hardly standards at all; they merely provided certain maximum limits of permissible bacteriological impurity. Revisions of 1925, 1942, 1946, and 1962, enlarged and refined the original standards to include requirements regarding source and protection of supply, standard tests to measure bacterial content, chemical and physical criteria, and greater scope of application. The criteria have often fluctuated in response to changing scientific knowledge.[34] Yet even in 1962 the standards were based on the premise that the raw water source would come from relatively unpolluted waters: "The water supply should be obtained from the most desirable source which is feasible, and effort should be made to prevent or control pollution of the source. If the source is not adequately protected by natural means, the supply shall be adequately protected by treatment."[35]

Morbidity and mortality data of acute human health effects have historically demonstrated whether or not public water supplies met the standards. But the obvious widespread, inadvertent re-use for potable supply and the murkiness surrounding chronic human health effects called for a major change in the standards.[36]

The Safe Drinking Water Act of 1974 for the first time established mandatory national minimal water standards. The Act instructs the EPA to set national standards for all contaminants considered harmful and to formulate temporary standards within three months with permanent standards to follow within four years. States are to enforce the standards, but if they neglect to do so, EPA can intervene. The Act allocates $156 million over three years to implement the provisions. Effective implementation could have far-reaching effects upon the future of direct potable re-use systems. If the new standards are comprehensive and restrictive, recycling systems could find it difficult to meet the standards economically.

*Water pollution control*. Compared to water supply, federal involvement in water pollution control comes at an even later date. The Water Pollution Control Act of 1948 culminated some fifty years of effort and gave the first legislative recognition of pollution as a national issue. Though recognizing

33. Dworsky, p. 105.

34. Ibid., pp. 87–88 ff.; Gilbert F. White, *Strategies of American Water Management* (Ann Arbor, University of Michigan Press, 1969), p. 64.

35. U.S. Department of Health, Education and Welfare, Public Health Service, *Public Health Service Drinking Water Standards, Revised 1962* (Washington, D.C., Government Printing Office, 1969), p. 2.

36. For a discussion of these issues in regard to direct potable re-use, see C. A. Hansen, "Standards for Drinking Water and Direct Re-use," *Water and Wastes Engineering*, 6 (1969), 44–45.

37. See U.S. Depart-
ment of the Interior,
Federal Water Pollution
Control Administration,
*Summary Report: Ad-
vanced Waste Treat-
ment, July 1964–July
1967* (Cincinnati,
FWPCA, 1968).

the primary responsibility of the states, the Act charged the Public Health
Service with developing comprehensive pollution abatement plans, established a Division of Water Pollution Control in the Public Health Service,
extended credit for the construction of municipal waste treatment plants,
provided a federal role for research and technical support, and authorized
$27 million (of which only $11.4 million was appropriated) in support over an
eight-year period.

By 1956 it was clear that only minimal progress had been achieved. The
Federal Water Pollution Control Act of that year extended and strengthened
the 1948 Act, and was the first legislation that authorized federal grants on a
large scale to assist states and municipalities to plan and build new municipal waste-treatment plants. The Act also provided increased technical assistance to states and an intensified research program, and it authorized the
collection and dissemination of basic data on water quality. Under the provi-
sions of this Act, some five billion dollars were appropriated and obligated
between 1957 and 1972. Subsequent amendments in 1961 increased the
grants for a waste-treament plant, constructed water pollution control laboratories in seven regions, and, perhaps most important for water re-use,
established the Advanced Waste Treatment Program, which supported important research and demonstration projects during the rest of the 1960's.[37]

Despite the progress growing out of this legislation, it did not keep pace
with the growing tide of wastewater and the scale of water pollution problems. The Water Quality Act of 1965 gave to states the opportunity to formulate water quality standards applicable to either streams or pollution sources
but failed to set precise quality criteria or settle the issue of how the clean-up
would be supported. Most effective, perhaps, was the implementation by the
Army Corps of Engineers in 1970 of the 1899 Rivers and Harbors Act, ban-
ning most dumping into navigable waterways.

But it was PL 92-500, the Federal Water Pollution Control Amendments of
1972, which provided the potential for far-reaching federal action. The principal thrust of the legislation was the point-source regulation of pollution
rather than regulation of the receiving body (for example, stream standards).
Moreover the Act provides clear goals, with a timetable for implementation.
By 1985 there would be a goal of *zero discharge* of all pollutants into navigable waterways. In the interim years of 1977 and 1983, industries and
municipalities should employ the *best practicable* pollution control technology (1977) and the *best available* technology (1983) respectively. For
municipalities this means secondary treatment of wastewater by 1977, more
advanced treatment, including water reclamation and re-use if practicable,
by 1983, and complete wastewater renovation by 1985. Congress then authorized some $18 billion for the first three years, and increased the federal
share of sewage treatment plant construction from 55 to 75 percent.

By 1975, however, it was clear that the program was lagging seriously.
Industry had made only 20 percent of the expenditures necessary to achieve
the 1977 goal, and the country as a whole was only 25 percent of the way to
the 1983 deadline. The federal goverment, which must bear the major share
of blame for this slow progress, had spent only $1.03 billion of the $18 billion

required. It is already clear that the follow-through is less ambitious than the legislation.

Even partial achievement of the legislation, however, could produce dramatic increases in water re-use by American cities over the next decade. In effect, the Act requires all cities to engage in water reclamation. Moreover, Section 208 of the Act requires regional planning for wastewater, including reclamation and re-use where feasible, in all metropolitan areas. This is potentially a powerful lever for widespread development of water re-use systems in American cities. In effect, water re-use as a means of increasing municipal water supply could easily slip through the back door; that is, cities will quickly find uses for the high-quality water produced by reclamation. New uses could easily defray the municipality's share of the costs as well as relieve pressures upon the water supply.

There are problems, of course. The legislation is keyed to technological capacity and the interpretation of such terms as "pollution," "practicable," and "feasible." President Nixon delayed the program by impounding six of the $11 billion authorized for 1973 and 1974. Nevertheless, by 1974 President Ford had begun to release this money. With the legislation and the funds beginning to become available, much will depend upon the federal agencies involved in water reclamation and re-use.

*EPA, The Army Corps of Engineers, and the land disposal controversy.* At the end of the 1960's and the beginning of the 1970's, two events came together which could structure the innovation process in water re-use for the remainder of the century. The first involved a water pollution planning project in the small city of Muskegon, Michigan. The County Planning Board in 1968 retained John Schaeffer, a geographer with the Urban Studies Center at The University of Chicago, as a consultant for the project. Schaeffer, the designer of Mount Trashmore, aware of the successful efforts of sewage application to land at Pennsylvania State University, took the lead in formulating a wastewater reclamation plan. The scheme involved utilization of municipal effluent from 14 municipalities and 200 industries in a reclamation system of aerated and storage lagoons, chlorination, subsequent spray irrigation on a 10,000-acre tract of land, and collection of the drainage for discharge to streams.[38]

Opposition from state environmental agencies held up the plan for a number of months. Eventually, project sponsors won an endorsement from the Federal Water Quality Administration (now the EPA). But the State Department of Public Health conducted a searching review and insisted on a number of changes in the original design. Eventually the project did win approval and was soon attracting national attention.[39]

The second event occurred during the same period. The Army Corps of Engineers, long involved in flood control, was seeking in the late 1960's to broaden its involvement in water resources development and hoping to gain a foothold in the newer areas of environmental planning and control. It was clear that its traditional role was in a declining industry. High officials in the Corps, concluding that future water resources problems would center on the

38. See Muskegon, Michigan, Muskegon County Board and Department of Public Works, *Engineering Feasibility Demonstration Study for Muskegon County Wastewater Treatment Irrigation System*, Water Pollution Control Series 11010 FMY (Washington, D.C., U.S. Department of Interior, Federal Water Quality Administration, 1970).

39. For a recent review of the Muskegon experience which suggests that a public relations firm should have been hired from the outset, see John C. Postlewait and Harry J. Knudsen, "Some Experiences in Land Acquisition for a Land Disposal System for Sewage Effluent," *Recycling Municipal Sludges and Effluents on Land*, ed. Environmental Protection Agency, et al., Proceedings of the Joint Conference on . . . (Champaign, Illinois, July 9–13, 1973), pp. 25–38.

40. U.S. Army Corps of Engineers, Office of the Chief of Engineers, *Wastewater Management Program: Study Procedure* (Washington, D.C., 1972).

41. Egelund, pp. 1471–73.

cities, explored a number of options, of which wastewater management was one, for Corps involvement. Cleaning up wastes, of course, would not only get the Corps into a lucrative planning area but also help refurbish its somewhat tarnished image.

In 1970, therefore, the Corps retained John Schaeffer as a science adviser to the general counsel to the Secretary of the Army. There had been little interest by the Corps in land-disposal methods of water re-use prior to Schaeffer's arrival, but the prospect of reclaiming the wastewater of the largest American cities in massive land-disposal projects must surely have whetted the appetite of professional engineers. And land disposal in the hands of the Corps was a totally different creature than the limited but effective projects created by the officials of small towns and cities earlier in the century.

In 1971 the Corps applied for and received congressional authorization to realign ongoing studies so as to include the development of planning alternatives for wastewater management in the four major metropolitan areas of Cleveland, Detroit, Chicago, and San Francisco, as well as York, Pennsylvania, and the Merrimack Basin in New England.

The objectives of the pilot wastewater feasibility studies were to produce:

(a) An overall wastewater management plan to meet the needs of 2020;
(b) A priced and evaluated portion of the plan that will meet the needs of 1990;
(c) A phased early action plan for the region to meet the needs of 1990;
(d) If appropriate, a proposal for congressional authorizations of selected elements of the early action plan.[40]

The Corps field offices prepared the feasibility study with the involvement of other state, regional, and local agencies. Each study purportedly considered both tertiary treatment plants and land-disposal strategies, but it was clear that the latter was the preferred choice. The plans, initiated in 1972 and to be completed in eighteen months, were in 1975 still incomplete for the San Francisco and Merrimack sites.

Despite a congressional stipulation, originally forced by the Office of Management and Budget, that the EPA was to participate jointly in the creation of these studies, it quickly turned into a one-agency show. EPA never played any major role in the program, and relations between the two agencies became very strained. William D. Ruckelshaus, then director of the EPA, protested that the studies were seriously biased in favor of land disposal over established technologies despite a lack of basic information and adverse economic considerations. In fact, Ruckelshaus attempted, unsuccessfully, to cut about $5 million from the funds allocated for the continuation of the study.[41] Substantial opposition also arose among environmental professionals in state, regional, and local governmental agencies.

A number of lessons came out of the pilot wastewater management program. The heavy emphasis upon land disposal—needed, it is argued, to stimulate serious planning with this option—produced a storm of hostile public reaction. In addition, although the Corps was able to demonstrate that land disposal can be a viable means of advanced waste treatment (given adequate pretreatment), the plans indicated that the magnitude of the system and the distance of the disposal site from the city are of crucial importance to the viability of land-disposal systems. Moreover, this technology encounters only firm disbelief among local and state public health officials, perhaps the single most influential group in water re-use decision-making. The problems in the plans were compounded by a Corps penchant for the grandiose and the spectacular at the price of sounder, less dramatic, designs. In view of these issues, it is not surprising that the plans have had little impact on wastewater and re-use programs in the cities studied.

Several other less immediate but more lasting impacts of this effort merit more attention. On the positive side, the program sensitized many local officials to the possibilities for water re-use and put them on notice that this is a coming innovation. It also undoubtedly helped to lay the groundwork for subsequent planning under Section 208 of PL 92-500. And the Corps also learned several things. Whatever the private opinions of high-order uses of reclaimed water, it is clear that the Corps has little or no capability in epidemiology, microbiology, and sanitary engineering. EPA holds the final cards on re-use in this area. Secondly, the Corps in its urban studies program, which currently involves water resources planning in some 35–36 cities and a budget of $11 million, has somewhat backed away from its emphasis upon land disposal of wastewater. It is unfortunate that the lessons cost a massive planning effort; careful reading of the water re-use record to date would have helped to avoid these pitfalls. More unfortunate, however, very little work was allocated to areas such as coordinated land use and water planning in multiple distribution systems, where the Corps would have had greater expertise.

The Environmental Protection Agency, moving more cautiously than the Corps, is presently laying the groundwork for future planned potable re-use systems. In Washington, D.C., at the Blue Plains Plant, it has instituted an extensive wastewater treatment program that will specifically test issues of treatment reliability. EPA has also inaugurated long-range studies examining organic concentrates and toxicological studies on animals. A major effort is under way, in short, to develop water renovation models with tested reliability for potable use when the need arises. This even extends to land-disposal technologies, although there is vehement resistance to its adoption among large elements of the Agency.

Much of the current work at EPA focuses, and with good reason, upon the upgrading of existing community water supplies in the United States. In 1974 the Agency launched a national study of contaminants in urban water supply, a logical background study for the implementation of the Safe Drinking Water Act. And the implementation of PL 92-500 will continue to occupy much of the attention of the Agency over the next decade. In addition, the Agency is deeply involved in the search for new water standards that will explicitly take account of water re-use in potable water supplies. These standards will undoubtedly be very stringent, and their exact form, together with the results of the national study of community water supplies, could shape profoundly the feasibility of direct potable re-use for urban water supplies and the practicability of alternative wastewater reclamation technologies.

Federal legislation and activity are very clear on one point—water re-use is upon us today. Even partial implementation of PL 92-500 will produce a tremendous growth in water re-use in American cities. And decisions within EPA over the next five years should largely determine the standards and eventual designs for the water re-use systems.

# Water Reclamation and Re-Use in Six Cities

David McCauley

**3**

The two preceding chapters have explored the nature of water re-use as a means of augmenting municipal water supply and examined the record of re-use activity to the present. They provide a meaningful general context for analyzing water reclamation and re-use innovations in specific situations prevailing in six cities that have adopted water re-use.

Several considerations guided the selection of these cities. First, given the scarcity of adoption of high-order uses of reclaimed water, it is important to include pioneers, which have afforded others an opportunity for empirical research. Lubbock in Texas and Santee and Antelope Valley (Lancaster) in California have had re-use efforts involving aquatic parks with lakes created from reclaimed wastewater. South Lake Tahoe, California, has the most advanced wastewater treatment plant and has recently added a re-use dimension. In Colorado Springs, Colorado, imaginative managers exploited the concept of "plural" or "hierarchical" supply discussed in Chapter 1. After field work was under way, researchers were fortunate enough to stumble upon an ongoing water crisis in which a community (San Angelo, Texas) was actively considering a planned, direct potable re-use system.

The city profiles that follow provide a brief inventory and description of the dimensions of the water-supply system relevant to any future decision concerning water reclamation for municipal use. Included are such variables as population changes, historical development of water supply, demand for water, water-supply and waste-treatment situations, previous experimentation (if any) with water re-use, and managerial attitudes toward re-use. Because the focus of this study is on the diffusion of an innovation (namely, urban water re-use systems), precedents and models for adoption are of particular interest. This chapter provides background information for the later more analytical chapters, which focus on obstacles to re-use adoption and on the overall process of innovation.

## Lubbock, Texas

Lubbock, on the West Texas High Plains, is a rapidly growing city in a semi-arid environment. The population (corporate limits) in 1940 was 31,850; by 1950 it had doubled to 71,400; by 1960 it was 128,691; and by 1970, 146,400. The city has firmly embraced the growth ethic and has a number of sectors, municipal and commercial, dedicated to this objective.

1. Approximately 350,000 acres were irrigated in 1964. The 1964 U.S. Census of Agriculture ranked Lubbock County second in the state of Texas for value of all crop sales.

2. Average mgd from 1940–1970 are:

| 1940 | 3.6 mgd |
|------|---------|
| 1950 | 9.3 mgd |
| 1960 | 18.2 mgd |
| 1965 | 25.5 mgd |
| 1970 | 26.0 mgd |

3. For an interesting, general discussion of this problem, see Walter Firey, *Man, Mind and Land* (Glencoe, Free Press, 1960). "Aquifer mining" is a problem throughout much of the Southwest.

4. "A Brief History of the Water Supply System for Lubbock, Texas," n.d. (mimeographed.)

Accompanying this population growth has been a rapid expansion of irrigated acreage in the area,[1] and Lubbock's economy reflects this development. Processing of cotton and cotton products has been a particularly strong element in this growth. Often regarded as the "Cottonseed Oil Capital of the World," Lubbock boasts a major production capacity. Grain, sorghums, and soybeans are also major crops, and Lubbock is a center for the regionally important livestock industry.

The city, in addition, has a rapidly growing heavy industry component. Major items include irrigation pipes and equipment, tractors, and heavy moving machinery. The processing of clothing and agricultural products is another element in the manufacturing picture. Lubbock is a major wholesaling center, but this does not greatly affect its water-supply situation.

Lubbock's water consumption has, of course, increased even faster than its population. The 1940 consumption of 3.6 mgd had increased by 1970 to 26 mgd.[2] Virtually all the agricultural users and most industries are self-supplied. The other large water consumers are, however, potential users of reclaimed wastewater. Even in food processing there are applications for water of less stringent quality requirements. In 1971 the new Southwest Public Service power plant, a private utility competing with the municipal power authority, contracted with the city to receive a maximum of 3.5 mgd of reclaimed wastewater for certain of its processes. Further expansion of reclaimed water use in this area could occur if water scarcity became critical.

Lubbock's water supply has traditionally come from wells in the Ogallala formation, a large water-bearing stratum underlying the region. The city owns 5,000 acres of water rights in and adjacent to the city, and by 1953 had acquired 2,000 additional acres of water rights in a nearby site. Competition for this "local" water is considerable, however, since agricultural, industrial, and domestic wells also draw on this source.

By 1957 the city had also acquired water rights to 75,000 acres in Bailey and Lamb counties which form the Sandhills Tract, a block 28 miles long and ranging from 2½ to 8½ miles in width. Pipeline systems and nearly 80 wells were installed, and by 1966 an estimated 18,000 acres of the Sandhills well field had been developed. Some managers expressed considerable concern, however, over the declining water table. Heavy pumping (municipal and agricultural—particularly in fields close to the city) and inadequate recharge constitute, in effect, "mining" of this aquifer, which is clearly a limited option.[3]

With completion in April 1967 of the Canadian River Project, Lubbock's water-supply capacity increased considerably. Lubbock was a member city in the Canadian River Municipal Water Authority (CRMWA), created by the Texas Legislature in 1953, which contracted in 1960 with the federal government to build the Canadian River Project. Lubbock receives 37 percent of the normal water supply available from the Authority. This figure includes the allotments of six small nearby communities for which Lubbock treats the Canadian River Water.

Lubbock's water treatment plant, completed in 1967, has a nominal capacity of 42 mgd. Since 1968 Lubbock has received 80 to 85 percent of its municipal water from the Authority.[4] Lubbock uses Canadian River Water to replenish during the winter its partially depleted local wells, treating the water before injecting it. During the summer, wells are used to supplement the CRMWA supply. The well water is chlorinated only; no additional treatment is necessary.

The Canadian River water is impounded in Lake Meredith near Amarillo, Texas, and reaches Lubbock through a 125-mile aqueduct. The switch from well water to Canadian River water (with different taste, odor, and color properties) created a minor political crisis in Amarillo, and was resolved only when the costs and difficulties of returning to and remaining exclusively on well water were outlined. In Lubbock the integration of Canadian River Water into the system involved only minor adjustments. There were some complaints but no organized opposition about certain properties of the water,[5] and managers did not see the incorporation as a serious problem.

The estimated safe yield for the total Lubbock sources is 88 mgd. With consumption averaging only 26 mgd in 1971, there is great flexibility in supply. Managers see present sources as adequate until 1995, at which time additional sources now in planning will be in service. Since those investigations were in a preliminary stage, they were not "public information."[6]

Lubbock is currently using all of its Canadian River Water allocation and will continue to do so, but this allotment is not expendable. Additional wells can be drilled in the sandhills field as they are needed. A perceived panacea for long-run supply exists in the Texas Water Plan, a scheme to bring East Texas or Mississippi River water to the West Texas area. This is some time off, however, in nearly all its aspects (planning, economics, and politics), but then, so too is any "shortage." Lubbock's "additional sources" will not be needed for some time.

Any projected shortage could be postponed even further, for there are a number of options in both allocation and demand management. Officials could, for example, transfer water from agricultural to municipal uses. Municipal uses enjoy legal priority in many states, institutions exist or can be created for these purposes,[7] and the economic value of the water is clearly higher in such uses.[8] The political problems are obvious, however, particularly in an area so closely tied to irrigated agriculture. Farms here are large, and farmers wield considerable local political power. Attempting close regulation of aquifer pumping would encounter the same problems.

Demand management through pricing changes is another possibility. Alteration of Lubbock's traditional declining block-rate structure could serve to discourage consumption, particularly by large users.[9] In any city a proposed change would involve political problems, but in Lubbock (and other semi-arid areas) these problems would be more complex. A water ethic seems to exist which invests water resource questions with more significance than they have in the East. This ethic would probably resist pricing changes. Discouraging consumption among large users may also be seen as contradicting the strong growth ethic. The net effect of both renders pricing changes less likely.

The City of Lubbock, under direction from the regional Texas Water Quality Board, recently upgraded its sewage treatment facilities. The former system involved passing wastewater through trickling filters (17-mgd capacity)

5. The total dissolved solids (TDS), for example, vary considerably. Canadian River Authority water averages approximately 1060 ppm for TDS, whereas that of Sandhills water averages 435 ppm.

6. "A Brief History."

7. L. M. Hartman and D. A. Seastone, "Alternative Institutions for Water Transfers: The Experience in Colorado and New Mexico," *Land Economics*, 34 (1963), 31–43.

8. See Nathaniel Wollman, *The Value of Water in Alternative Uses* (Albuquerque, University of New Mexico Press, 1962).

9. Residential and Commercial Rates:

First
3,000 gals. $3.25 (min.)
Next
7,000 gals. .50 per 100 gals.
40,000 gals. .40 per 100 gals.
200,000 gals. .38 per 100 gals.
250,000 gals. .34 per 100 gals.

10. The farmer who was currently using the effluent had a contract with the city. The Canyon Lakes proponents were also concerned about how the powerplant sale would affect their goals. See the *Lubbock-Avalanche Journal* (April 25, 1968).

11. For general statements, see Henry Graeser, "Dallas Wastewater Reclamation Studies," *Journal of the American Waterworks Association*, 63 (1971), 634–640. See also Daniel Cannon, "Industrial Reuse of Water: An Opportunity for the West," *Water*, ed. Roma McNickle, Western Resources Conference, 1963 (Boulder, University of Colorado Press, 1964), pp. 69–78.

12. Three phases are identified in the engineers' report (Lake 7 is Phase II, and Lake 8 is Phase III). Freese, Nichols, and Endress, Inc., Consulting Engineers, *Feasibility Report on the Canyon Lakes Project* (Fort Worth, Texas 1969), p. 12.6.

13. Other indirect objectives were mentioned. See Lubbock Planning Department, *An Expanding Lubbock . . . Reclaimed Water for a Growing City* (1968).

14. See Lubbock Planning Department, *Canyon Lakes Project: Goal, Program, Policies* (n.d.).

and then spreading the effluent on city and privately owned farmland. This process irrigates nonedible crops and prevents the direct discharge of effluent into the Brazos River. Effluent-fed springs do, however, augment the flow of the Brazos which is impounded downstream to form the eutrophic Buffalo Spring Lakes. The completion of a 12-mgd-capacity activated sludge system added to existing treatment facilities. This effluent, of generally good quality except for rather high nitrate concentrations, is also used for land disposal. The city intends to use it in its water re-use scheme.

Some Lubbock officials anticipate a number of uses for the reclaimed water: irrigation of parks and golf courses, replacement of potable water with reclaimed water in irrigated agriculture, satisfaction of several industrial water requirements, and certain recreational uses. Of these, the last two merit particular attention here.

Reclaimed wastewater for the Southwestern Public Service company's generating plant will provide a sector (i.e., industrial water) of traditional municipal supply. As stated earlier, the maximum amount to be delivered will be 2½ mgd. Such provision normally involves a simple administrative decision but is complicated in this case by client competition for the effluent.[10] This type of re-use would not involve a major innovation for, as Chapter 2 notes, a number of cities (Odessa, Amarillo, Big Spring) in the region and elsewhere practice wastewater re-use for industry.[11]

More revolutionary is the aquatic park with recreational lakes, a facility that would not have existed in Lubbock had it not been for wastewater reclamation. Water re-use not only permitted the creation of such a facility, but also extended municipal water resources services beyond the traditional domestic, industrial, and commercial sectors.

The Lubbock Urban Planning Department was the central agency in the formulation of the Canyon Lakes Project, which originally envisioned two phases. Phase I was to involve six small lakes within the city limits, and Phase II Lakes 7 and 8 (the largest), which would be outside the city limits.[12] Lakes 7 and particularly 8 were to provide virtually all of the recreation; power boating and swimming were to be reserved for Lake 8. The Phase I lakes were to be primarily ornamental, with no body-contact recreation planned. Financing arrangements for the two phases also differed. Lubbock would finance Phase I and regional funding would create Phase II, although it was hoped that both phases would receive some federal support.

The Urban Planning Department believed that the Canyon Lakes Project had many desirable objectives: local and regional recreation interest, local and regional growth, water supply storage, neighborhood development, and land-use expansion and improvement.[13] Indeed the proponents of the Canyon Lakes Project vigorously espoused "the long-range recovery type goals" of the scheme and succeeded in having it (along with parks and urban renewal projects) included in a tornado-recovery bond issue. Voters approved the Canyon Lakes scheme by a resounding two to one.

The interest in wastewater re-use was due primarily to its contribution to the recreational-ornamental lakes, the central element in the plan, but the provision of recreation per se was clearly secondary to the enhancement of land values and the ornamental and land-use benefits.[14]

The reclaimed wastewater lakes were tightly linked with a land-planning scheme.[15] The Planning and Parks Departments envisioned an important recreational potential for Yellow-Horse Canyon, even though it had become a home for many undesirable land use activities (auto salvage, some feedlot operations). The Planning Department, then, in a *tour de force*, packaged the clean-up need for the Canyon and the required effluent storage into the Canyon Lakes Project.

The Planning Department treated the Council to a slide presentation outlining the possibilities of the Canyon Lakes Project and included provisional data on cost and benefits. The Council reacted favorably and commissioned a consulting engineering firm to conduct an economic-engineering feasibility study of the project. (An earlier study by a different firm recommending the use of effluent in the Southwestern Power Plant was a factor in the Planning Department's suggestion of using reclaimed water for the Canyon Lakes.) The consultants returned a basically favorable report, and the city appointed a full-time project coordinator.

The Planning Department, which reported a favorable public reaction to its program and slide show, expressed particular surprise that few people objected to the use of reclaimed water. Indeed, no interest groups opposed the project, and a number expressed outright support for it. Public reaction, while not unanimous, was reported as good.[16]

The general level of interest in water-resource questions in Lubbock (and in the other semi-arid sites) is high, particularly among municipal officials but also among the general public. For some elected officials the Canyon Lakes Project played an important part in the campaign platforms of some elected officials.[17] Moreover, as stated earlier, the scheme received public support in a bond referendum.

During the course of the public information program, officials visited Santee to find out what of the Santee experience might apply to Lubbock. Among the lessons learned from Santee was the necessity for an honest, public-information strategy outlining fully the benefits and costs. There was less interest in the bacteriological and virological data relevant to swimming at Santee, because Lubbock did not plan for swimming in Phase I of the project.

Although Planning Department officials expressed concern over accidental ingestion and possible odor and appearance problems in the Phase I Lakes, the fear of contamination by storm runoff from the streets, a hazard identified by the consultants' report,[18] was the primary reason for disallowing swimming. They also expressed caution about swimming until "we see how the water behaves in the lakes." The small size of these lakes was not amenable to other significant water-based recreation.

With a favorable feasibility study in hand, project officials sought financial assistance from state and federal units. They approached the Bureau of Outdoor Recreation (BOR) and the Army Corps of Engineers, as well as state and federal congressmen. Financing arrangements were insecure at the time, and after some initial support they collapsed.[19]

The first four dams were completed early in 1976. Since the original conception, however, lack of financial support and the location of an interstate

15. Editorial, *Lubbock-Avalanche Journal* (February 13, 1968).

16. "Canyon Lakes Plan Needs More Study," *Lubbock-Avalanche Journal* (May 11, 1970).

17. Political advertisement, *Lubbock-Avalanche Journal* (May 13, 1968), pp. 14–15.

18. Freese, Nichols, and Endress, p. 12.2.

19. *Lubbock-Avalanche Journal* (April 29, 1971).

20.  The population figures are drawn from the U.S. Bureau of the Census, *General Population Characteristics: Texas*, 1970 Census of Population, OC (1)–B45.

21.  Letter from Thomas L. Koderitz, Water Superintendent, City of San Angelo (September 7, 1972).

22.  There were, presumably, other objectives. One can observe how the U.S. Army Corps of Engineers and Bureau of Reclamation divided among themselves the responsibility for "protecting" San Angelo. West Texas is a jurisdictional boundary zone.

23.  Koderitz letter.

highway have caused two of the lakes of Phase I to be dropped. In addition, the inability to find financial support for Lakes 7 and 8 of Phase II, the major recreational parts of the plan, resulted in their deletion. As a result the project is now strictly an ornamental and land-use facility.

## San Angelo, Texas

San Angelo in 1970 had a population of 63,884, an increase of 5,000 over the 1960 population of 58,815.[20] Unlike Lubbock, the San Angelo municipal government does not promote rapid growth. There are few individual large water users in San Angelo, the major user being a meat-packing plant. Municipal authorities have generally not encouraged large water users to locate in San Angelo. The increased stress on the water system is, in fact, one deterrent to a policy of urban growth.

By 1971 the water-supply situation in San Angelo was very serious. The city "reached a low in supply in April of 1971 when we had only three months' supply in all the reservoirs."[21] The lack of runoff, due to the unusually low rainfall, caused serious concern among municipal officials. As a result, San Angelo began to investigate the feasibility of using reclaimed municipal wastewater for domestic consumption. The crisis situation was the culmination of a series of water-supply problems. A brief outline of the water system and of these problems follows.

The total average water use in San Angelo in 1971 was 10–11 mgd. There is a major peaking problem in the consumption—from a low of 5 mgd during the winter to a high of 28 mgd for summer peaks. Although a filter capacity of 25 mgd limits the safe yield of the system, this yield could be easily increased. Reservoir storage is capable of holding up to a five-year supply.

Up to 1971 San Angelo had depended entirely upon surface runoff for its water supply. Prior to the construction of its first reservoir, pumps at the confluence of the North Concho and South Concho rivers provided the city's water. Low flows in these rivers led in 1930 to the construction on the South Concho River of the Nasworthy Dam (12,000-acre-feet capacity). Construction of other city reservoirs alleviated water supply and flooding problems (the city had experienced "severe flooding" in 1882, 1936, and 1957–59).

In 1953 the U.S. Army Corps of Engineers constructed the San Angelo Reservoir (120,000-acre-foot supply storage and approximately 300,000-acre-foot flood storage) on the North Concho River. In 1960 there were 120,000 acre-feet of storage in the San Angelo Reservoir. But this was a high point, and the water levels began to decline thereafter. By 1970 the reservoir was dry.

The 1957–59 flood of the South Concho River produced a decision to impose additional control on this river.[22] By 1963 the Twin Buttes Reservoir, a multipurpose project of the Bureau of Reclamation which included municipal water supply, irrigation storage, and flood control among its objectives, had been completed. Since its construction, however, the reservoir has held little water; by 1970 it, too, stood "empty with barely the dead storage wet."[23]

Although the rainfall during the middle and late 1960's was not low, the runoff was. Some speculated that changed watershed characteristics (e.g., rapid expansion of mesquite trees) were responsible for the decrease, while others maintained that the rainfall pattern (i.e., infrequency of heavy runoff producing downpours) was the culprit. The runoff stored in these reservoirs was subject to the high evaporation rates characteristic of this area. In San Angelo the large surface area that was exposed aggravated the evaporation problem. Because of the broad and low relief, the reservoirs were wide and shallow.

The Colorado River Municipal Water Authority had completed the E. V. Spence Reservoir on the Colorado River by early 1970. The city of San Angelo installed a thirty-mile pipeline to the reservoir site in 1970, but for the first year and one half the reservoir failed to catch sufficient water.

Water-use restrictions were imposed from 1969 to August of 1971, as they had been during earlier periods (e.g., the drought of 1952–53). By restricting the hours of water use and discouraging the use of water for certain activities (e.g., car washing), municipal officials exhorted users to conserve water. Temporary pricing schedules designed to discourage water use were reported to have had "little effect since people were already conserving."[24] Public awareness of, and concern over, the water supply situation was very high.[25]

Quality deficiencies (although safety was not a factor) also arose during this crisis period. With the depletion of the reservoirs through usage and evaporation, the total dissolved solids (TDS) began to climb and reached in March 1971 a high figure of 1430 ppm. Dilution, of course, proceeded rapidly when the rains came. It was even necessary on several occasions to treat and use the runoff from the city streets to augment the municipal water supply. This undesirable measure met with opposition, and although municipal and public health officials expressed displeasure, they reluctantly consented to it as a necessary step. As an emergency measure, this was avoided, of course, whenever possible. During the shortage, at times when the perceived quality of the water was poor, many people bought bottled water and complained to municipal officials.

Because there were no other exploitable surface alternatives, the city also began investigating underground sources. Discussion centered on two options. The first was to explore a nearby aquifer (south of Christoval, Texas) where, although the water was of good quality, the amount available was limited. Moreover, it was locally recharged, and officials feared that drought conditions affecting San Angelo's surface supplies would also affect this source.

The second option involved a source more distant (50 miles from the city in McCulloch County) and deeper (at least 3,000 feet of pumping would be necessary). The supply was expected to be large but this was uncertain. Investigations through test wells suggested a figure of 20 mgd for 50 years "without undue depletion." But if the quantity were questionable, it was at least certain that the cost would be high (a bond issue for at least $12 million was mentioned). This decision was under consideration in July 1971, when the research team was in San Angelo. Since that time, the city has pur-

24. Interview with John Williams, former Water Superintendent, City of San Angelo (July 1971).

25. It was still high in July of 1971. Many people seemed convinced that the rains were only a temporary reprieve from a water supply condition which could only be alleviated by new (and large) water sources.

26. Interview with John Williams, former Water Superintendent, City of San Angelo (July 1971).

27. Some municipal income, a small amount, was earned from the sale of animal feed, which was irrigated with treated wastewater on the city farm.

28. A complication existed here. The effluent before ground-spreading did not meet state standards, but after land infiltration and before entering a stream (where state jurisdiction normally begins), it did. The city decided to improve its treatment level even though this question remained.

29. The city was experimenting with a pilot project similar to the proposed physical-chemical process.

30. The water would be mixed with Concho River water. This was emphasized more from the point of ensuring homogeneity of the product rather than as a hazard control measure.

chased the water rights to over 37,000 acres in McCulloch County and undertaken a well-drilling program.

Also under consideration during this period was planned, direct potable re-use. Since it was necessary to upgrade the wastewater treatment facilities, some argued that "so long as the treatment increments were planned we might as well go all the way toward water reclamation."[26] The extant system involved only primary treatment followed by application to the ground at the city farm.[27]

San Angelo already intended to increase the capacity of this system, but pressure from state water-quality standards made it necessary to improve the quality of the effluent as well.[28] The city requested of its consulting engineers four feasibility plans, one of which involved treatment levels adequate for reclamation. The engineers proposed a physical-chemical plant, and some interaction with Calgon Corporation, a producer of this type of system, ensued.[29]

Then the rains came. In April 1971, with only three months' supply left in Lake Nasworthy, heavy rains increased total system storage by 10,000 acre-feet. The sense of crisis abated somewhat. In August 1971 another heavy rain period provided 100,000 acre-feet in Twin Buttes Reservoir, and the Colorado River put 110,000 acre-feet of storage in Spence Reservoir (which San Angelo shares with four other Texas communities). San Angelo Reservoir, however, remained dry.

With the increase in runoff there was a decrease in interest in wastewater reclamation for water supply. The city backed off its temporary wells in the nearby aquifer at Christoval but proceeded with a well-drilling program in McCulloch. What did the water managers feel were the various costs and benefits associated with direct potable re-use at that time, and how did the water-management system respond to it?

It is important to point out here that the public never received from the municipal government any proposal of re-use for domestic consumption. The City Commission (mayor, council, city manager) and some municipal employees discussed it as an option but never actually endorsed it. They did, however, commission the consultants to do a feasibility study, and re-use did have its advocates in the system. System members possessed high information levels about water re-use, and during the period of discussion some members met with Environmental Protection Agency (EPA) officials. Some City Commission personnel even visited the facilities at Lake Tahoe.

Wastewater reclamation in San Angelo was viewed in the context of drinking, and river mixing was the method discussed.[30] But commission members expressed a number of reservations concerning quantity, quality, hazard, cost considerations, and difficulty with Public Health approval. In part because of these, McCulloch County wells were the preferred alternative.

The proposed wastewater reclamation program would produce only 5 mgd of high-quality effluent. Officials felt that this amount was an insufficient long-term water supply. Some expressed reservations about the build-up of TDS in the re-use process, particularly since the initial raw water was poor in this regard. Some saw viruses as a serious block to adopting direct potable re-use.

But there were other reservations. Some of the treatment processes were new and uncertain. There was insufficient lead time to design, construct, and operate the plant, and test its effluent sufficiently to rely on water reclamation. Some felt that the essential Department of Public Health (DPH) approval would prove difficult to obtain. Also, some regarded cost as a problem. The incremental cost beyond what would have been necessary to meet state standards[31] would not have been that great, however, particularly set against the $12 million necessary for McCulloch County wells. Used as an emergency supply during droughts, the re-use system could also delay new increments in capacity. This is exactly the type of system discussed below in Chapter 5.

The City Commission decided to investigate re-use possibilities. For the Commission to propose and endorse water reclamation for drinking to the public, a number of things would have been necessary. Endorsement by local professionals—people the public "knew and trusted"—would be essential. During the re-use discussions, support from local college professors proved very helpful. Managers viewed their support and that of doctors as essential. Extensive testing of the facilities and interaction with state and federal officials would also, of course, be necessary. Yet the commission still felt that this was "an issue for the public to decide," and they would have submitted the issue to a referendum. The reasons were twofold: (1) the project would have to be financed, and (2) an expression of public feeling would be desirable. In talks with EPA and other governmental officials, some discussion centered on San Angelo as a demonstration city. The city encouraged the label, because demonstration status would mean financial aid and free testing. Incorporation of the effluent in the supply system would, however, still be submitted to the public for approval.

The political strategy would have been to "present the issue to the public honestly, in terms that they can understand, and let them know the costs and risks." This, it was felt, would reduce the political risks. There was considerable emphasis on being prepared to answer all questions; one should minimize "the number of times you have to say I don't know." Managers felt the demonstration effect was important and that "much of the groundwork had been laid at Tahoe and Santee." They cited various aspects of public information strategy: use of radio and newspapers, television programs, public hearings, lectures to community groups and schools, and use of films. The water manager in his appearances at public gatherings carried a bottle of reclaimed water with him and drank it. This gesture was felt to be highly effective.

Water managers did expect some opposition. Those "whose minds can never be changed" and who will vigorously oppose water re-use can best be handled, political managers felt, by creating occasions (television shows, public hearings) where technical experts can directly contradict their arguments. The object of a public information campaign is to convince the middle group, the largest, whose votes can mean passage of the referendum. Virulent opposition groups are containable and can be counteracted by supportive groups. In San Angelo the League of Women Voters, for example, showed interest in supporting the re-use idea.[32]

31. The Water Superintendent, a strong re-use proponent, emphasized that he would not have recommended wastewater reclamation if it were not already necessary to increase treatment levels.

32. In other study sites, the managers usually sought and obtained the support of the League of Women Voters for reclamation and re-use.

33. Letters to the Editor, *San Angelo Times* (March 1971).

34. There was no conflict over fluoridation in San Angelo. The water is naturally fluoridated (0.4 to 0.7 mg/L).

35. It could, of course, serve as an emergency supply. If the high yields hoped for in McCulloch County are obtained, however, its use is highly unlikely.

36. Private communication with Thomas Koderitz, Water Superintendent, February 12, 1975.

37. For one point of view on the relative distribution of benefits of this plan, see G. Marine, "The California Water Plan, The Most Expensive Faucet in the World," *Ramparts* (May 1970), 34–41.

Even though the City Commission did not propose direct potable re-use, information that it was under discussion did reach the public, producing some early pro and con opinion. A number of letters were published in the newspapers,[33] and some of the issues raised (pure water being ruined by municipal action, everyone must drink it) recalled those debated in fluoridation controversies.[34] It was too early in the re-use decision process for interest group activity to have reached any significant level, but one suspects that should the issue (for drinking) be raised again, this activity would become significant.

Following the rains which alleviated the 1971 crisis, the dry years returned. By July 1974 the Twin Buttes reservoir was again down to 50,000 acre-feet. At that point the city stopped supplying all irrigators. By October of the same year rains again not only filled the reservoir but went to flood stage. By early 1975 the San Angelo Reservoir, dry for many years, was one third full, and Twin Buttes was accumulating 500 acre-feet in flood storage per day. With 10 wells completed in McCulloch County, another 7½ mgd had been added to the system's safe yield (although the pipeline still must be built). With this rosier water situation, the city has decided upon an activated sludge sewage treatment plant (and not tertiary treatment) and has submitted a grant request to EPA. Clearly the re-use prospect is all but dead for the immediate future,[35] and, according to the water superintendent, "most people seem to have forgotten about reclamation. The only people who seemed much disappointed were the college professors."[36]

## Santee, California

The Santee County Water District, a public agency, provides water and sewer services for a rapidly growing suburban district in the San Diego metropolitan area. Although the population in Santee was only 500 in 1956, the water district served 14,500 people in 1970, and the projected population in the area for the year 2000 is 100,000 people. Santee is primarily a bedroom community. Industry is mostly of the light variety, and the industrial-commercial sector of the water use is not appreciable. Most of the water use and most of the effluent are therefore domestic.

Though in a dry area, Santee has no water shortage. "Water scarcity" and "water consciousness" are, however, elements of a public image. The Santee Water and Sewer District distributes and retails Colorado River water. While the Central Arizona Project may have caused some supply concern in the Santee area, the controversial California Water Plan has improved the gross supply picture.[37] Santee may receive Northern California water (better quality, 200–300 ppm, TDS) directly or it may receive more Colorado River water (800–850 ppm, TDS) through trade-offs with upstate communities. The Metropolitan Water District will not run dry.

Santee's water supply may be secure, but the price for this security is considerable and the cost of imported domestic water is steadily rising. The

TABLE 3.1

*Wholesale Purchase Price of Water in Santee, 1959–1973*[38]

| Year | $/Acre-Foot | $/Mil. gal. |
| --- | --- | --- |
| 1959 | 19.00 | 58.33 |
| 1967 | 36.25 | 111.28 |
| 1973 (est.) | 75.00 | 230.25 |

Santee Water District paid a wholesale price of $53 per acre-foot in August of 1971 for California Water Plan water for municipal use.

The sewage situation is more local, at least for the present. In 1959 Santee chose to eschew the Metropolitan Sewer District. In 1958 the community passed a bond issue for a sewage treatment plant. By 1959 a 1.5-mgd unit was put into operation which included primary and secondary units plus one or two oxidation ponds on land donated by a major local housing and golf-course developer. The same year, because of (1) the previous investment in the new plant,[39] (2) an aversion on the part of a major actor to single use and then ocean disposal, (3) the expense of entering the metropolitan system, (4) an evolving idea of and commitment to reuse, and (5) an interest on the part of a client (golf course) in the effluent, Santee declined to join the new Metropolitan Sewer District. If the Santee plant had not already been built, the community would almost certainly have joined the metropolitan system, eliminating what has ensued at Santee. Ultimately, as will become evident, it was forced to join the system in any case.

After the plant was built in 1959, Santee tightened water-quality standards somewhat and initiated the practice of moving the effluent from pond to pond. A fairly good effluent was being produced, and the district manager, the major actor in this innovation, became committed to using it. He recognized the golf course irrigation, through well withdrawal at the golf course rather than direct pumping,[40] as an obvious use. He looked for others. The focus here, of course, is on recreational lakes, whose example has been widely emulated elsewhere.

Geologically, Sycamore Canyon, in which the Santee Lakes are situated, is particularly suitable for the recreational lakes program. A layer of impermeable clay underlies (at an average of 10–15 feet) the sands and gravels. This plus the slope ensure lateral movement and easy collection of the water. The clay also seals the lakes, so that water loss through percolation is negligible. Technological replication of these geological advantages could, of course, be provided elsewhere but at an expense.

Also avoided was the cost of major lake-site excavation.[41] The developer-golf course owner donated the site for the lakes in exchange for water rights priority. Both the district manager and developer continually emphasized the potential utility of the site for reclaimed lakes and thereby influenced the direction of the excavation to preserve trees and other sites (now islands in the lakes). For this reason certain of the lakes are rectangular

38. Figures from E. W. Houser, "Santee Project Continues to Show the Way," *Water and Wastes Engineering*, 8 (May 1970), 41.

39. This investment would not have been reimbursed, nor would the facilities have been used.

40. Direct pumping was being considered in 1974.

41. The District carried out some small amount of excavation.

42. Houser, p. 44.

43. These and related points are treated at considerable length in Leonard A. Steven, "Every Drop Counts," *National Civic Review*, 56 (March 1967), 142–147, 155.

44. See John Merrell, et al., *The Santee Recreation Project: Santee, California—Final Report*. (Washington, D.C., U.S. Department of Public Health, 1967).

and lacking in islands, whereas others (excavated after the re-use potential was realized) are styled. It is extremely significant that this process involved no major cost to the district. Because the excavation would have proceeded anyway, Santee practiced land, as well as water, reclamation.

Some major actors insist that the lakes program would have occurred even if the sites had not "been excavated for them"; they argue that the costs would not have been prohibitive. Tucson and Antelope Valley (and Lubbock to some extent) have excavated their sites, but only on the heels of the Santee example. In the early stages of an innovation, initial obstacles loom menacingly; one suspects that the extant excavation was an important consideration.

Also important in the Santee case were the significant contributions of money for studies and facilities from governmental agencies. In 1966 the District received approval of a grant under Federal Water Pollution Control Administration Public Law 660 to aid in the construction of a new 5-mgd primary and secondary conventional treatment plant which would lead directly into a tertiary system. The cost for this system was $2,040,000.

In December 1966 the District received an $800,000 demonstration grant from the Federal Water Pollution Control Administration to support the cost of constructing a 2.0-mgd tertiary unit and two years of its operation. In this system the removal of nitrates and phosphates through biological and chemical processes, respectively, precedes filtration.[42]

Santee, therefore, had many of its facilities subsidized. The receipt of these grants probably stemmed from the fact that this research was already under way in the District. The research also provided the district with some highly qualified personnel and aided in the establishment of national and international links for the District. This special attention, particularly with its funding and personnel support, will be available to some other early innovators but probably not to the same degree.

An existing social network strengthened the project organization. Two major actors, the district manager and developer, had been friends before and the developer (and land donator) was a client for the reclaimed water. Such strong bonds at the outset probably strengthened the commitment to complete the project. Other good fortunes (e.g., the accidental moving to San Diego of a prominent virologist and the presence of some progressive health officers) also aided the project.[43] Some of these people (e.g., a research project director)[44] had extensive knowledge of the problems and prospects of wastewater reclamation and furnished valuable technical aid to the Santee County Water District manager, who lacked such expertise. This is not to contend that these things cannot happen in other re-use sites but simply to underscore their occurrence at Santee and their roles in the innovation process.

Some people at Santee did know about the use of reclaimed water in Golden Gate Park in San Francisco, the total re-use project at the Camp Pendleton Marine Base (operating since 1942), and other projects, so the idea itself was not original. Realization in the Santee context grew under pressure from state standards, out of an effort to raise the quality of the effluent and to provide cheap water for a golf course. The first step involved the plant and an oxidation pond, but by 1961 the effluent was being moved from lake to lake in a three-lake chain, and project managers already realized its potential for recreation. But they needed approval from at least three sources—the Board of Directors of the Water and Sewer District, the general public, and the regulatory agencies (particularly the health agencies)—in order to continue. The latter two sources were the most important. Throughout the whole Santee innovation the Board of Directors was generally supportive and approved measures which satisfied the public and the technical agencies. A major factor in this support was strong confidence in the quality and reliability of the District's staff.

The process of informing the public had been going on as the idea was forming. Certain members of the district exploited many occasions— community club meetings, media, and informal gatherings—to inform people and sound out reactions. In 1961 the lakes (then three or four, now six) were improved by some landscaping, stocking of fish, and the provision of picnic tables, but they were surrounded by a chain link fence that restricted the public. The district manager felt that this would be a more effective way to build up public support than persuasion by verbal information. He also felt that strong public support would aid in dealing with the regulatory agencies. The strategy served at least two functions: (1) to show the public what could be done, and (2) to generate enough support to convince the regulatory agencies to confront the re-use problem so that standards could be developed.

Public support for and interest in the lakes did indeed develop. Public acceptance at Santee seemed to involve no problems.[45] With each increment in use, the public expressed more interest. In some cases public enthusiasm exceeded the managers'; one morning, before the facility was officially approved for any recreation, fifty people were found swimming in the lakes.[46] Support and confidence may also have been increased by the publicized encounters with regulatory agencies. Because each possible increment in use met with increased doubt, opposition, and sometimes outright refusal, people may have felt assured when a new step finally did win approval. This would seem to have been the case in Santee where interaction with the regulatory agencies was a major component of the innovation process.

Among the agencies directly interested in the Santee Project were the California Regional Water Quality Control Board Number 9, the State Water Resources Agency, the State Fish and Game Agency, the U.S. Public Health Service, and those of the State of California and the County of San Diego. Of these units the public health agencies played by far the major role and were concerned with the most substantive issues (especially viruses and health hazards). The basic pattern of district-agency interaction was a conflict-cooperation nexus: in some areas the health agencies might obstruct, in others they might be too lax; in Santee the combination of a progressive yet cautious orientation was the major thrust.

The Santee Water and Sewer District wanted to open the area for recreation (nonbody contact) in 1961, but the San Diego County Health Department, citing insufficient treatment and public health risk, refused permission.

45. With recreation lakes, unlike drinking, a selection process occurs. People who strongly disapprove simply do not go to the lakes. There is not this option with reclaimed water for domestic supply and bottled water presents an added burden.

46. Wesley Marx, *Man and His Environment: Waste* (New York, Harper and Row, 1971), p. 118. Marx includes a general treatment of Santee in this book.

**David McCauley**

Figure 3.1: Views of Water Re-Use in Santee.

At this time Santee was moving the water through three lakes. The Public Health officer would issue his approval when Santee met two conditions: extra treatment and the initiation of a virus analysis. The District added the further step of pumping up the valley and filtering through gravels and began the major Santee virus studies. The area was opened for nonbody-contact recreation after health personnel were satisfied with the initial treatment and the results of the virus study.

The conflict-cooperation pattern manifested itself on two other major occasions. In the fishing program, fishing for fun was the first step followed, but only after a period was "take-home-and-eat" fishing allowed. Another controversy involved swimming (in a pool, not in the lakes). Before approving this step, the District and the regulators interacted intensely. They did not advance beyond a closely supervised, controlled program[47] until the Public Health officials were satisfied that no significant hazard existed. The use of reclaimed water in the pool continued for approximately six years, until 1968, when, with the termination of federal funding, the high cost of maintaining the esthetics terminated the program. A chemical reaction caused the water in the pool to turn yellow-brown, an unappealing color for reclaimed water. Colorado River water now fills the pool. Water quality or hazards were not the issue, nor were they the central problems in the proposed lake swimming; turbidity, reduced visibility, and the consequent susceptibility to drowning-related lawsuits were the major deterrents.

The major conflict was between the "cautious incrementalism" of the regulatory agencies and the enthusiasm of the innovation champions in the District. In retrospect, most of the major Santee actors agree that, frustrating as it was to go so slow, the proper pattern did emerge. The conflict-cooperation pattern existed in Santee in part because the major parties respected one another and a number of communications channels were kept open. In a more rancorous dispute, or in cases involving mutually low levels of respect for technical and personal attributes, a different pattern could have emerged.

Wastewater reclamation at Santee was important primarily for its recreational lakes. The lakes were a regional as well as local attraction, an observation made by a small group that opposed Santee's decision to provide facilities for nonresident visitors.[48] There was a general benefit to users because of the scarcity of this type of recreational facility in an arid area. There was also a more specific value to the developers whose homes were situated near (particularly with a view of) the lakes; land values near the lakes increased, and the houses became substantially more expensive. Many observers also reported an intangible community benefit emanating from the lakes.[49] At one time the very positive experience with the lakes probably would have made a move to potable use easier to implement.[50]

Of course, more than recreation benefited from water reclamation at Santee. The golf course irrigation used and still uses reclaimed water. By 1974, however, there were problems with pipes clogging, and although the course owner continued to use reclaimed water, he did so reluctantly. The reclaimed water was also used to irrigate a tree farm and a Little League park.

Additional uses were discussed while the reclamation project was in full operation. A nearby community college considered landscape irrigation

47. This involved restricting to twenty-five the number using the pool for one-half hour and then admitting a new group. Names, addresses, and health records were carefully kept. No waterborne disease was reported.

48. There was a small and poorly organized group that objected to "pouring money into the recreational lakes." No community conflict developed; the community generally supports and is proud of its lakes.

49. There is, for example, a Santee Lakes Festival and a Miss Santee Lakes. Whether these benefits are tangible or intangible is a matter of opinion.

50. Public acceptance levels for drinking were higher in Santee than in the other sites (see Chapter 7).

51. The city of Ocean-side, California, has discussed reclamation for potable use. Whittier Narrows and Camp Pendleton, California, as indicated above, practice ground-water recharge for general municipal withdrawal. For a fuller list and discussion, see Chapter 2.

52. During the "innovative period" the district manager, a former skin-diver and golf pro, was very interested in recreation.

53. This is not to say that Santee is the first instance of re-use for recreation (Golden Gate Park and Camp Pendleton were earlier), but it is the major development to which other managers interested in re-use for recreation almost invariably point.

54. See John Parkhurst, "A Plan for Water Reuse, A Report Prepared for the Directors of the County Sanitation Districts of Los Angeles County" (July 1963). (Mimeographed.)

55. Editorial, *Antelope Valley Press*, (August 30, 1962).

water at one time, and a potential market existed for freeway landscape irrigation. Another lake was also considered in 1971. The failure to find other customers for the reclaimed water was a major cause of the eventual demise of the project.

At Santee the major period of innovation with wastewater reclamation and re-use is over. Because of an abundant alternative supply and the total control and high cost necessary for achieving potable re-use, the long-run prospect was not favorable.[51] But in any case the project soon ran into more immediate difficulties, for by 1971 governmental funding for programs at Santee had largely dried up, and district revenues from water and sewer service appeared inadequate to carry on innovative reclamation programs with their expensive treatment processes. The District management apparently came to regard recreation as a distinctly secondary function and concentrated more directly on its water and sewer functions.[52] Then problems emerged with the operation of the re-use system itself. Solids and dissolved solids became increasingly troublesome in water quality control. The sludge problem caused difficulties. Although the oxidation ponds initially appeared to require no sludge removal, in actuality the sludge build-up over time became detrimental to their operation. In any event the County Health Department and the Regional Water Quality Control Board, claiming fish and game damage and an emerging problem with nutrients in the water which were promoting the growth of dense brush and mosquitoes, issued a cease-and-desist order for discharging the effluent into the river. Confronted by operational problems, unable to find customers for the reclaimed water, and facing political tensions, Santee finally capitulated and joined the San Diego Metropolitan District. One of the important water re-use innovations in America, and one which still serves as a model, died without fanfare.[53]

## Antelope Valley (Lancaster, California)

The Antelope Valley Project is one of the re-use schemes in a comprehensive program of wastewater reclamation carried on by the Los Angeles County Sanitation Districts.[54] It is the only project in the program which uses reclaimed water for recreational lakes, although the overall program includes a wide range of uses—from prevention of sea water intrusion to ground-water recharge for general municipal withdrawal.

The previous chapter described the Los Angeles County Water Reuse Plan. The Antelope Valley Project was designated as a goal in 1962, and by January 1972 the lakes were full. The Antelope Valley Project involves approximately twenty-six acres of lake surface in three individual but interconnected lakes. The lakes project emulates the Santee model, visited by project originators as a source of information and, it seems, "inspiration."[55]

A total domestic wastewater flow of 4 mgd from Lancaster, California, undergoes secondary treatment in oxidation ponds. Approximately .5 mgd of the oxidation pond effluent receives tertiary treatment and is allocated to the lakes to compensate for evaporation. Some of the water is also withdrawn from the lakes for irrigation purposes and to aid in circulation of lake water. The tertiary treatment process consists of coagulation with alum, sedimentation, and passage through dual media filters. During coagulation, phosphates are precipitated along with most of the algae carried over from the oxidation ponds, while filtration removes the remaining algae. The reclaimed water is then chlorinated and conveyed through a five-mile pipeline to a new aquatic park.

The cost of the treatment plant, park, and lakes was approximately $2,240,000, of which Los Angeles County paid 48 percent and state and federal agencies the remaining 52 percent. During the 1960's proponents of the project conducted considerable research into treatment processes, fish studies, and bacteriological and virological questions. This information appears elsewhere;[56] the emphasis here is on the intended reclamation application and certain social and political aspects of the innovation.

56. John Lambie, *Waste Water Reclamation Project for Antelope Valley Area: Final Report* (August 1968).

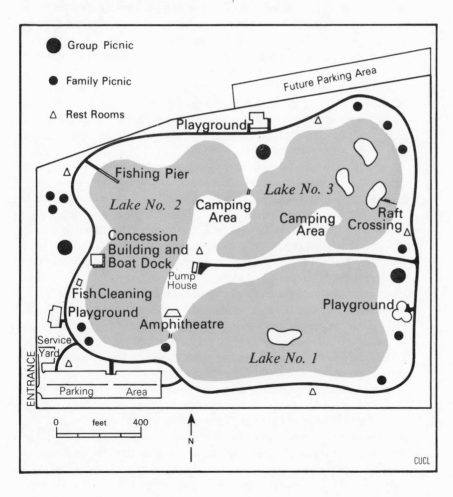

Figure 3.2: The Antelope Valley Wastewater Reclamation Project.

57. *Los Angeles Times* (June 21, 1967), Part II, p. 6.

58. *Desert Spectator*, 2 (July–August 1963), 14.

59. *Antelope Valley Press*, (December 2, 1965), p. 4.

60. Ibid. (February 24, 1963), p. 7.

61. Editorial, ibid. (August 30, 1962).

62. *Los Angeles Times* (June 21, 1967), Part II, p. 6.

63. The use is not restricted to dry areas, as the recreational lakes at Michigan State University in East Lansing, Michigan, attest. It is, however, more popular in drier areas, as the lake experiments at Antelope Valley, Tucson, Lake Tahoe (Indian Creek Reservoir), and Lubbock indicate.

The Antelope Valley area is on the southwestern edge of the Mojave Desert. Falling ground-water tables, a lack of water recreation, and insufficient water for desired economic expansion have enlisted many enthusiastic supporters in the project. "We live in an arid region," said the President of the local Chamber of Commerce." Lack of sufficient water has prevented many companies from establishing plants here. When the water becomes available, we expect a tremendous influx of industry and a surge in the local economy."[57] Images of the recreational potential were equally euphoric.[58]

The context of the organization of the Antelope Valley Project embodies two main elements. The first was the evolving commitment on the part of the Los Angeles County Sanitation Districts (Lancaster is located in District No. 14) to wastewater reclamation. A second, and extremely important, influence was the Santee experience. It was after a trip to the Santee site and a preliminary evaluation of its suitability that the Antelope Valley Project was born.[59] Local supporters and clients for reclamation were not hard to find. In fact, a conflict arose between a local water agency and the county over who would develop and manage the reclaimed water.[60]

Client support was important in the Antelope Valley Project (as in Santee and Lubbock). Each development received general public approval for its contribution to the goals of recreation and growth, but in each case a more specific set of clients for the reclaimed water was instrumental in moving the project along. In Antelope Valley support involved County officials as well as local residents. One official, a Los Angeles County supervisor, was a particularly strong supporter.[61]

The expected benefits from the project were recreational opportunities, industrial water supply, park irrigation, regional growth, and the enhancement of land values. The current treatment-plant output of reclaimed water is 4 mgd, and the treatment facilities have a design capacity of 4.5 mgd. Because of the significant loss of flow to evaporation and limitations in plant capacity, only slight expansion in the program is possible.

A larger public information program accompanied the development of the Antelope Valley Project idea. Meetings, lectures to community groups, and frequent articles in newspapers all played a role in the project's development. There was little adverse public reaction, although some older people objected to using reclaimed water.[62] Managers attributed the favorable public reaction to the effectiveness of the public information program, favorable endorsements from local doctors and public health professionals, and the particular desirability of an aquatic park in this dry area.

Intended uses at the Antelope Valley recreation area include picnicking, overnight camping, fishing, and boating. No water-contact activities, such as wading or swimming, are planned. The water managers would prefer to "see how the water behaves in the lake," and gauge public reaction to the actual reclaimed lakes before proceeding further. There is presently, however, no intention of allowing swimming. As sites like Antelope Valley and Canyon Lakes (in Lubbock, Texas) follow the Santee example, the resistance encountered by proponents of reclaimed lakes at other possible sites will lessen. Using reclaimed water for recreational-ornamental lakes in dry areas appears to be a particularly suitable innovation.[63]

## Lake Tahoe, California

"Keep Tahoe Blue" was the slogan. It was the symbol for a process that produced the most advanced waste-treatment facility in the United States—that of the South Tahoe Public Utility District (P.U.D.) Unquestionably a major technical achievement,[64] it is significant primarily as an operational, full-scale, continuously working plant using wastewater treatment processes already known.[65] No less an achievement at South Lake Tahoe was the political action that brought the facility into existence.[66]

The permanent population of South Lake Tahoe is approximately 20,000, but during the tourist season it increases to 75,000 or more. Since the same situation exists in Tahoe City, during heavy tourist periods the population in the Tahoe Basin may exceed 200,000 people. Lake Tahoe, because of the frequently exposed but always shallow granite bedrock, is a lake of exceptional clarity and is known nationally and internationally for its beauty and recreational facilities. Fears over pollution of Lake Tahoe resulted in the Keep Tahoe Blue movement.

One of the major objectives of the drive was to press for high-level treatment and export of the wastewater from the Tahoe Basin. The current treatment plant has a capacity of 7.5 mgd. The plant includes primary, secondary (activated sludge), and full-capacity tertiary systems that provide chemical

64. For a complete statement on the Lake Tahoe operation, see Russell and Gordon Culp, *Advanced Wastewater Treatment* (New York, Van Nostrand, 1971).

65. Interview with Russell Culp, Director, South Tahoe Public Utilities Department (August 1971).

66. We restrict this to South Lake Tahoe. At the northern end of the lake, the Tahoe City area, considerably less success has been achieved.

Figure 3.3: Water Reclamation at South Lake Tahoe. Reprinted by permission of Cornell, Howland, Hayes, and Merryfield, Clair A. Hill and Associates.

67. Russell Culp and Harlan Moyer, "Wastewater Reclamation and Export at South Tahoe," *Civil Engineering* 41 (1969), 38.

68. A good study of the attitudes of the decision-makers in the Tahoe Basin is available in Edmond Costantini and Kenneth Hanf, *Environmental Concern and Political Elites: A Study of Perceptions, Backgrounds and Attitudes*, University of California, Davis, Institute of Governmental Affairs, Research Reports, 21 (Davis, California, 1971), p. 40.

69. For an early (1964) technical perception of Tahoe's disposal alternatives, see F. H. Ludwige, E. Kazmierczak, and R. Carter, "Waste Disposal and the Future of Lake Tahoe," *Journal of the Sanitary Engineering Division*, ASCE, SA3, 90 (1964), 27–51.

70. Interview with Mr. Hubert Bruns, Chairman, Alpine County Board of Supervisors (August 1971).

treatment and phosphate removal, nitrogen removal, mixed-media filtration, activated carbon absorption, and disinfection.

The total investment in the collection system, pump stations, treatment plant, and export facilities is approximately $37.5 million, of which South Lake Tahoe residents have provided about $24.5 million; federal grants from the Economic Development Administration and the Environmental Protection Agency, some $6 million; the U.S. Forest Service (for sewer services on their lands), $2 million; and the State of California, $5 million.[67]

Because of the dependence of the local tourist-based businesses on "a clean Lake Tahoe," the treatment and export objective, even at this high cost, received considerable local support. Enthusiasm for these water-quality goals was not, however, restricted to the tourist industry.[68]

The specific mandate to export the wastewater from the basin stemmed from a Federal Water Pollution Control Administration Conference held at the lake in 1966. All wastewater was to be exported by April 1, 1968. The new South Lake Tahoe system went into operation at the end of March 1968, and wastewater export to Alpine County began.[69] Other export destinations had been discarded by South Tahoe P.U.D. because of political opposition at the site. The initial public reaction in Alpine County was generally negative, and the Alpine County Board of Supervisors was divided on the issue.[70] People were not enthusiastic at the prospect of receiving "somebody else's effluent," but the combination of standards imposed by State and Federal agencies and by Alpine County ensured that the effluent would be of high quality and could benefit Alpine County. Opposition gradually decreased with the spread of information and positive experience with the water. Alpine County residents have come to look favorably, if at all, on Indian Creek Reservoir, the body of reclaimed water.

The water-quality standards imposed by Alpine County were more stringent than the federal-state regulations, and it was these county standards which the Tahoe plant had to meet. The treatment level probably would have been lower were it not for this pressure. Some view the intentionally high standards as a tactic that would render it unlikely, given the technical problems and high costs of meeting these requirements, that South Lake Tahoe P.U.D. would use Alpine County as an export site. Even if this is not true, Alpine County's demand that the effluent be of high quality seems justified, and the standards prevent slippage from this demand. Another reason for the high quality standards was that officials planned to use this effluent eventually for water-supply needs in the Tahoe Basin.

With a high growth rate, the Tahoe Basin can anticipate a projected water shortage within ten to fifteen years. Aware of this possible shortage, the Public Utility District's Board of Directors recommended high treatment levels in 1961 for two reasons: to ensure the use of water in the basin twenty-five years later, and to demonstrate to regulatory agencies that a high-quality effluent could be continuously produced.

The re-use possibilities in the Tahoe Basin appear, however, to be limited. Drinking the water is not under consideration nor should it be with Lake Tahoe so close at hand. There are no large nonpotable users in the area, nor is agricultural use a significant factor. Golf-course irrigation and cooling uses (presumably air conditioners) are possibilities, but these would involve relatively small amounts of water. Ground-water recharge is not a possibility because of geological restrictions.

Regardless of how probable one regards the projected water shortage to be, it appears that the political impetus that brought about the high-quality effluent may prevent its re-use in the basin. A political campaign obtained approval for the construction of facilities to treat and export a high-quality effluent to protect a lake. The requirements for export, however, have produced effluent of a quality so high that the threat to the lake has probably vanished and the need for export diminished. Ironically enough, the intention to re-use this effluent within the basin will probably run afoul of the same factor that resulted in its production—namely, the desire to "Keep Tahoe Blue," even as the concern over water shortage or limited pressure for re-use grows.

After the fulfillment of contract obligations, the considerable pumping costs could be saved by discharging the renovated wastewater into the lake or a tributary of the lake. The prospect, however, elicits a strong reaction from "Keep Tahoe Blue" supporters. This does not necessarily mean that the situation cannot be changed, but suggests rather that efforts to re-use the effluent in the Tahoe Basin may meet with considerable opposition despite universal acknowledgment of the high quality of the effluent. The prevailing context is "clear and blue" Lake Tahoe.

Although use of reclaimed water for recreation is not feasible in the Tahoe Basin, it is in Alpine County. The present recreational use of Indian Creek Reservoir is limited. All-year trout fishing for edible take-home catches is one activity at Indian Creek, but the number of fishermen is small.[71] The Bureau of Land Management has responsibility for the operation of day and overnight camping units, boat ramps, and certain other facilities.

Of greater importance are water-contact sports at Indian Creek Reservoir. Boating entails a ten-horsepower restriction on motor size, ostensibly imposed because large motors were regarded as incompatible with trout fishing. Reservoir size, not fear of hazards associated with the water quality, prevented this form of water-contact sport. In fact, the lake water has been approved for all water-contact sports, including swimming. The reservoir would undoubtedly be used more heavily for this purpose were it not for its muddy bottom. A sand beach is one of the facilities yet to be developed.

Indian Creek is also a source of irrigation water. As supplementary irrigation in Alpine County for the past six years, it has encouraged some small expansion of cropland. It is used in pasture and fodder irrigation for livestock, even though California permits higher uses on agricultural products. While the income generated from this use of water has been negligible, it has contributed to public acceptance of the project in Alpine County. Irrigation possibilities may also have helped the initial acceptance of the Tahoe water (some of the supporters on the Alpine County Board of Supervisors were also irrigators). Aware of the complications and problems of western water law, the South Tahoe P.U.D. has exercised considerable caution to retain ownership of this water and prevent loss of control over it.

So, although the P.U.D. produces an effluent which is cleared for all

71. For a brief treatment of this point (but which does not include Indian Creek), see Robert Hallock and Charles Ziebell, "Feasibility of Sport Fishery in Tertiary Treated Wastewater," *Journal of the Water Pollution Control Federation*, 42 (1970), 1656–65.

**David McCauley**

72. All supply and use figures are from Black and Veatch, Consulting Engineers, *Report on Water Distribution System for Colorado Springs, Colorado, 1969 Revision* (Kansas City, Mo., 1969).

73. See the discussion of the rationale behind this strategy in Ernest Flack, "Urban Water: Multiple Use Concepts," *Journal of the American Water Works Association*, 63 (1971) 644–646.

body-contact sports, administrative difficulties currently restrict its use to a very low level. High quality allows for effluent application in a wide range of uses, but potential political problems in the basin and a small range of re-use possibilities have virtually stifled most potential uses. It is ironic that one of the highest quality effluents in the United States has found so few uses.

## Colorado Springs, Colorado

Colorado Springs in 1950 had a population of approximately 45,000 people. It had increased to 70,000 by 1960, and to approximately 120,000 by 1968. The number of customers served by the City's Water Division for that period, and with projected demand, are shown in Table 3.2.[72]

The average daily water use over the same period is: 1966, 31 mgd; 1970, 38 mgd; 1980, 52 mgd; and 1990, 64 mgd.

Colorado Springs, in an area of the country widely regarded as environmentally attractive, is growing rapidly. The municipal government in 1971 appeared to embrace firmly the growth ethic, and the water management system was strongly committed to providing the foundations for such growth. "The growth of the city," said one manager, "depends entirely upon water supply." The city allocated (in 1968) over 37 percent for residential uses. Industrial supply requirements received approximately 20 percent of the total, leaving 22 percent for municipal needs and 21 percent for the military (major defense-related institutions such as Fort Carson are located in this area). Although the specific figures differ from year to year, the approximate respective shares of the allocation will probably not change radically. This range of large users has major implications for the re-use experience at Colorado Springs.

The history of water-supply acquisition in Colorado Springs is familiar in the western context. From development of local ground and surface supplies, the city has participated in large and expensive transmontane diversions in competition with other metropolitan areas.[73] By 1934 the south slope of Pike's Peak and other local sources had been almost completely developed (including the tapping of local ground water). The firm yield was estimated at 4 billion gallons annually, and in 1939 the water system had no bonded debt. After World War II, however, the city's population began its rapid expansion and the acquisition of major new supply sources started to surge.

Colorado Springs began receiving a share in the Northfield system diversion in 1949, and financing began on the Blue River System in 1950. These projects aimed to provide 5 to 6 billion gallons annually. In the early 1960's, Colorado Springs and Aurora, Colorado jointly began development of the $60-million Homestake Project, whose annual yield was estimated at 12.5 billion gallons, with the first phase providing 4.5 billion. During this period, storage and distribution improvements were included in the massive acquisition program. The potential yield of the system in 1970 stood at 22.5 billion gallons annually, while the maximum annual consumption to that date had

TABLE 3.2

*Growth in Colorado Springs Water Division*

| Year | No. of Customers |
| --- | --- |
| 1960 | 83,400 |
| 1968 | 138,500 |
| 1980 | 273,000 |
| 1990 | 394,000 |

been only 13 billion gallons. Supply acquisition continued; procedures for integrating into the system the Frying-pan-Arkansas Project (3 billion gallons annual yield) also began.

The pattern in water supply management is clear—source expansion with a preference for clean mountain supplies has dominated. One might expect this pattern, exacerbated by competition from other metropolitan areas engaged in similar pursuits, to persist as long as new expansion sources were available. But the constriction of alternatives of this type, because of the development of, or claims on, the existing exploitable sources, has alerted water managers to the limitations of this pattern. And, hope eternal (e.g., the diversion to this area of Alaskan or Canadian water) notwithstanding, managers have come to acknowledge the utility of water reclamation and re-use within this context.[74] Indeed, re-use may be "the last water hole,"[75] at least in Colorado Springs.

The sewage treatment system consists now of primary and secondary treatment with a small (3-mgd) additional tertiary unit. The secondary system treats approximately 13 mgd as an average, although minimum and maximum figures range from 7 mgd to 32 mgd. The secondary system includes trickling filters, secondary clarification, and chlorination. After a 30-minute detention time, the effluent is released into the receiving stream, Fountain Creek, much of which is used for irrigation (a minimum of 7 mgd must, by law, be received by downstream users).

Approximately 5 mgd of treated effluent are pumped to the nonpotable-water station, where wastewater receives additional treatment by processing through two pressure sand filters to remove any humus remaining from previous treatment. It is subsequently pumped to gravity-flow storage reservoirs and then distributed to large users. The tertiary unit, which involves treatment by chemical coagulation followed by passage through activated carbon filters to remove both suspended and dissolved matter, provides reclaimed water for a power plant and for nearby irrigation users.

Though wastewater reclamation did improve water quality, the major perceived advantages to reclamation lay in its water-supply potential. The projected failure of traditional strategies also enhanced the role of water re-use.[76] A major goal was to supply nonpotable reclaimed water to users with some lower quality requirements, thereby releasing potable water for other higher-order uses in the system. This is, of course, the embodiment of the plural water-supply concept. A similar but more specific objective would provide reclaimed water to those water-dependent activities which would other-

74. It has also been recognized by the framers of water compacts where in the recent ones it has been stipulated that the diverted waters should be "used, re-used, and used again." Interview with James Phillips, Sanitation Superintendent, Colorado Springs.

75. Colorado Springs, Colorado, Department of Public Utilities (Sewer Division), "City of Colorado Springs Sewage Treatment System" (n.d.), p. 9: "Therefore when no more diversions can be made and the ground water supply is unstable, the sewage treatment plant again becomes an answer, since it deals in a water which does not wear out, but has only become temporarily not acceptable for normal use."

76. "After citing the water projects that have been integrated in the system, the question is asked, 'But is this still enough? The answer is no!' Where then is the last water hole that can be used to make 'enough' come true? Actually, the answer to this question is the sewage treatment plant." Ibid.

wise be excluded (e.g., certain irrigators, a golf course, cemeteries) or disadvantaged if a low-cost water were not available. Delivery of nonpotable water costs customers one third of the cost of potable supply.

The thrust, then, has been to provide an increasing segment of potable users with nonpotable supplies, thereby extending the system's capacity. The re-use idea here is a rational one which enjoys proper handling in an area and organization which plans on a major scale for water supply. It embodies in many ways the ideal type of water re-use discussed in Chapter 1. Note, however, that this rather tight net of economic efficiency is internal to a context whose economic basis (a vigorous competition for new supplies) is questionable. This affects the economic efficiency of reclamation and re-use.

Early examples of wastewater re-use in the area were at Fort Carson and the Broadmoor Recreation Area. They were known to the sewer department manager in the late 1950's. He disapproved of the idea of carrying water over the mountains, using it once, and then just flushing it. He was the man who proposed the sale of effluent to the Highway Department in 1957. During that year the State Highway Department approached the city of Colorado Springs for potable water for center-strip irrigation. The city rejected this proposal but offered sewage effluent, conditioned upon public health approval.

Public health approval permitted installation of a nonpotable line, but the Highway Department never exploited it. Drought conditions which had existed for a few years preceding 1957 produced two unexpected but needy customers—Colorado College and the Kissing Camels golf course—for the nonpotable line. A similar set of conditions prevailed in 1962 when citywide rationing forced the water department to shut off the potable water to dying city parks and a public golf course. The sewer department, in the process of installing a nonpotable line to serve the golf course, picked up other unexpected customers (cemetery, parks). Since that time other nonpotable users, mostly irrigators and industrial water users, have tied into the system.

Many of the initial hookups, then, were accidental, and drought conditions helped to create the initial impetus for the system's expansion. Now the program is one of *planned* replacement of potable with nonpotable users as opportunities arise.

Re-use in Colorado Springs exists within a context characterized by high levels of interest in water-supply questions and a strong commitment to their solution. Both water managers (technical and political) and the general public espouse this ethic, and their respective preferences rationalize and reinforce one another. Awareness of re-use relates to the projected supply deficiency, and managers argue that reclamation will enjoy further enhancement as the projected deficiences loom closer and the cost of water rises. Although the managers felt that these factors would increase the acceptability of reclaimed water for drinking, they feared psychological rejection and had no intention of moving to direct potable re-use in the near future. Indeed, with the planned replacement program, the rationale for this is steadily undercut.

There are problems with managerial acceptance. The water department approves of the nonpotable system and supply augmentation through wastewater reclamation, but the sewer department manages the nonpotable system. The water department balks at the notion of direct potable re-use, or cutting the reclaimed water into the general municipal water supply. The two departments informally agreed upon a ten-year experimental period for testing the reclamation and re-use process. The experiment has at least three objectives: (1) to determine the soundness of the treatment technology, (2) to observe whether the process can provide a continuous, reliable supply of water, and (3) to identify whatever other problems are inherent in this process.

Although water-cost increases, perceived shortage, and experience with re-use activities would probably guarantee public approval of direct potable re-use, the managers still do not endorse it. In addition to the psychological resistances ascribed to the general public, managers also foresaw (not without insight, as subsequent chapters will show) the prospect of a battle with Public Health officials. Neither confrontation seemed desirable to the managers (or the researchers).

The major actor in the re-use innovation, the Sewer Department Superintendent, did feel that current water-reclamation activity was having a positive impact on public acceptance. Although there had been some negative reaction to speeches and newspaper articles about re-use, it had come from older (approximately 50–70 years of age), more conservative people who were few in number and not organized in their opposition. The major reaction, managerial and public, to the re-use experiences in Colorado Springs has been very supportive.

No problems, managers felt, would confront lower-order uses of reclaimed water, but one had to tread very carefully with higher-order uses. In spite of the Santee model, managers were cautious, though still basically supportive, about recreational lakes. As for drinking, they were clearly opposed.

# Part II. Potential Obstacles to the Diffusion of Water Re-Use Systems

# Health Hazards Associated with Re-Use for Potable Supply

John T. Reynolds

4

There is little question that potential public health hazards can be associated with the potable use of treated municipal and industrial wastewaters. If one defines health hazards to mean "any conditions, devices or practices in the water supply system and its operation which create, or may create, a danger to the health and well-being of the water consumer,"[1] the weight of the evidence suggests serious uncertainties as to the availability now and in the short-term future of adequately monitored fail-safe systems guaranteed to remove or inactivate the full range of harmful chemical and biological agents presumed to be in municipal wastewaters. These agents include the wide variety of pathogenic organisms and toxic chemicals found in the community producing the sewage. Some of these agents produce acute, reasonably well-understood effects—typhoid fever, amoebic dysentery, infectious hepatitis, methemoglobinemia, etc. Others may be associated with more subtle, poorly understood, multifactorial effects—cancer, circulatory disorders, birth defects, etc.

Because the full range of potential hazards and their relative import is not clear, there is no consensus among competent authorities as to acceptable standards and/or criteria to be applied during the evaluation of potable supplies when that water is produced from municipal sewage.[2] Nevertheless, wastewater re-use to augment water supplies has been, is, and will be a planning factor in a growing number of communities.

For many reasons—geographic, economic, and political—water managers have tried the indirect use of reclaimed wastewater as a source of supply. According to the report of a World Health Organization (WHO) meeting of experts in 1973, "existing natural sources of drinking water in several parts of the world already contain industrial and municipal wastewaters in proportions that may approach 100% in periods of low flow."[3] There are estimates that in the United States "one-third of the nation's population currently takes water from streams which, on the average, have one gallon of 30 that has been used upstream and returned to the stream after use. In extreme cases, one gallon out of five has been used previously."[4]

It is fair to assume that, in the above cases, only a small fraction of the wastewater used had received advanced waste treatment before being introduced into the receiving natural systems. It is even more probable that a significant fraction had not received adequate secondary treatment. Rather, the "magic" of natural purification, buttressed by "careful" treatment-plant operation, was depended upon to protect the public health. However, the increasing demand for water and the concomitant increase in the volume of

1. U.S. Department of Health, Education, and Welfare, Public Health Service, *Public Health Service Drinking Water Standards: Revised 1962*, (P.H.S. Bulletins, 956 (Washington, D.C., Government Printing Office, 1969), p. 2.

2. World Health Organization, *Reuse of Effluents: Methods of Wastewater Treatment and Health Safeguards*, WHO Technical Reports, 517 (Geneva, 1973), p. 52.

3. Ibid., p. 45.

4. Jerome Gavis, *Waste Water Reuse*, U.S. National Water Commission Reports, NWC-EES-71-003, July 1971 (Washington, D.C., 1971), p. 6.

5.  U.S. Environmental Protection Agency, *Preliminary Assessment of Suspected Carcinogens in Drinking Water: Interim Report to Congress (with Appendices)* (Washington, D.C., U.S. Environmental Protection Agency, 1975), pp. 25–30.

6.  Ibid., pp. 26–27. See also World Health Organization (note 2, above), p. 26, and James M. Symons, et al., *National Organics Survey for Halogenated Organics in Drinking Water* (Cincinnati, EPA Water Supply Research Laboratory, 1975).

7.  Environmental Protection Agency (note 5, above), pp. 9–10.

8.  Ibid., pp. 6–8.

9.  Daniel A. Okun, "Alternatives in Water Supply," *Journal of the American Water Works Association*, 61 (1969), 216–217. See also Environmental Protection Agency, p. 18.

10.  James H. McDermott, "Virus Problems in Water Supplies," *Water and Sewage Works*, 122 (1975), 71–73, 76–78.

11.  Harold W. Wolfe and Steven E. Esmond, "Water Quality for Potable Reuse of Waste Water," *Water and Sewage Works*, 121 (1974), 48–54.

12.  World Health Organization, p. 16.

13.  Ibid., p. 52.

14.  U.S. Public Health Service, Bureau of Water Hygiene, *Community Water Supply Study: Significance of National Findings* (Washington, D.C., 1970).

wastewater to be processed have effectively reduced the time and dilution factors needed for the natural purification systems—integral parts of indirect water re-use strategies—to operate reliably. Furthermore, the diverse array of organic chemicals associated with the growth of modern agriculture, continuing industrial development, and accelerated urban growth characteristic of the mid-twentieth century are now present in wastewater.[5] Many of these are not completely removed by conventional waste treatment processes—sedimentation, filtration, and biological oxidation. Some are not readily degradable by the biological agents operating in natural or man-made treatment processes. In some cases, treatment may lead to the production of new compounds very different from the parent compounds.[6] As a result a growing number of organics are found in both raw[7] and finished[8] water supplies when they are sought. And since some of them are known to be toxic, carcinogenic, teratogenic, or mutagenic,[9] a new class of potential public health hazards—perhaps not adequately taken into account by existing drinking-water criteria and standards applied during treatment and surveillance practices—is beginning to attract considerable attention.

The unease about existing water criteria and standards is further increased by the growing awareness of the possible epidemiological significance of other ubiquitous agents present in both raw and finished drinking water supplies—for example, viruses,[10] and inorganics such as sodium.[11]

Nevertheless, as Chapter 2 notes at length, several well-publicized instances of direct re-use for potable supply have occurred.[12] A 1956 drought emergency forced Chanute, Kansas, to re-use a secondary effluent after passage through standard water treatment. Windhoek, Namibia, met its water needs for an extended period in the late 1960's by mixing the highly treated effluent from a sophisticated advanced waste-treatment process with wastewater from a natural source.

More intentional re-use for potable supply, direct and indirect, may occur in the near future. As of 1973, planning or research on the re-use of wastewater for drinking was going on in at least nine countries: India, Iran, Israel, the Netherlands, Singapore, South Africa, Thailand, the United Kingdom, and the United States.[13]

No untoward acute effects on the public health have been reported in either case of direct re-use and, presumably, because the experiences with direct use have been recent and of short duration, no long-term chronic effects have been noted. The broad-scale acute and chronic effects of exposure to general supplies dependent on indirect re-use have yet to be fully ascertained: in many cases the still prevalent water-borne disease outbreaks are assignable to breakdowns in water-treatment plant practices and inadequate surveillance,[14] and the comprehensive epidemiological studies of population groups with histories of long-term exposure to re-use have just begun.

Nevertheless, it appears to be the consensus among health authorities that "a high proportion of existing water supplies are so heavily contaminated with municipal and industrial wastes as to make the health risks associated with the drinking of water derived from such sources only slightly different from those involved in the direct re-use of waste water." And what significance is attached to those risks? "Drinking water should preferably come from a clean supply." But if an adequate "clean" supply is not available, "communities should make every effort to conserve (clean) water so that there is sufficient for potable purposes. Waste water re-use for all other purposes should come first."[15]

What prompts such caution? A partial list would include:

(a) recognition that present standards, plant operation, and surveillance practices are inadequate for the fail-safe operation of the systems deemed necessary when potable supplies are developed from municipal and industrial wastes;

(b) recognition that the essential toxicological and epidemiological information needed for the development of new criteria and standards to be used in quality control in such situations is not available.

A few examples may illustrate these problems.

## Drinking Water Standards

Existing standards for finished drinking water are based on the assumption that water for human consumption is to be drawn from protected ground water and surface supplies. If the most feasible source is not adequately protected by natural means, the wastewaters that may eventually be used as the raw water intake should be subjected to appropriate waste treatment. The level of waste treatment to be achieved is such that standard water treatment will result in the "production of a safe, clean, potable, aesthetically pleasing and acceptable supply which meets the limits of Drinking Water Standards after treatment." Standard water treatment may include coagulation (less than 50 ppm aluminum, ferric sulfate, or copperas with alkali addition as necessary but without coagulant aids or activated carbon), sedimentation (6 hours or less), rapid sand filtration (3 gal/sq. ft./min/ or less) and disinfection with chlorine (without consideration to form of chlorine residuals).[16]

The U.S. Public Health Service Drinking Water Standards, augmented by the recently released Environmental Protection Agency Interim Primary Drinking Water Standards, list thirty-three parameters to be used for assessing the acceptability of finished drinking-water supplies.[17] The WHO Drinking Water Standards contain a few more parameters which may serve as grounds for judging a supply. Neither of these widely used sets of standards establishes limits for all of the potentially toxic inorganisms that may be found in municipal and industrial wastes. Neither list includes more than a few of the synthetic organic compounds that must be presumed to be present in many municipal, agricultural, and industrial wastewaters, including many that are known to be health hazards. Both base their assessment of the microbiological health hazard on the detection and enumeration of bacterial surrogates, coliforms, supplemented by turbidity and residual-free chlorine requirements, for the varied bacterial, fungal, protozoan, helminth,

15. World Health Organization, p. 25.

16. National Technical Advisory Committee on Water Quality Criteria, *Report of the Committee on Water Quality Criteria* (Washington, D.C., Government Printing Office, 1972), p. 18.

17. Environmental Protection Agency, "Interim Primary Drinking Water Standards," *Federal Register*, 40 (1975), 11, 990–98.

18. Ibid., 11,995.

19. L. J. McCabe, "Health Aspects of Reusing Wastewater for Potable Purposes," Paper presented at the American Water Works Association Meeting of June 1975; William N. Long and Frank A. Bell, Jr., "Health Factors and Reused Waters," *Journal of the American Water Works Association*, 64 (1972), 220–225; World Health Organization, pp. 25–26.

20. Environmental Protection Agency (note 5, above).

21. Symons, et al.

22. Ibid.

23. EPA (note 5, above), p. 17.

24. Ibid., pp. 19–23.

25. Wolfe and Esmond, pp. 49–50.

26. U.S. Environmental Protection Agency, *Federal Register*, 11,992.

27. O. C. Liu, *Effect of Chlorination on Human Enteric Virus in Partially Treated Water from Potomac Estuary*, Northeastern Water Hygiene Laboratory Progress Reports November 1971, (Washington, D.C., EPA, 1971), pp. 5–13; American Society of Civil Engineers, Sanitary Engineering Division, Committee on Environmental Quality Management, "Engineering Evaluation of Virus Hazard in Water," *Journal of the Sanitary Engineering Division. Proceedings of the American Society of Civil Engineers*, ASCE, 96, SA1 (1970), 111–161.

and viral pathogens that could be present in municipal wastewaters.[18] Therefore, most of the detailed studies of the chronic health effects associated with the ever-increasing number of chemicals currently found in raw waters used after processing for general supply have yet to be completed. Yet the inadequacies of existing drinking-water standards are clear to a number of observers.[19]

Studies have demonstrated the presence of chemical carcinogens in river water and in treated municipal drinking water.[20] There is also evidence that the normal practice of disinfecting drinking water by chlorination, particularly if the water contains significant organic loadings, results in the production of a variety of chlorinated hydrocarbons some of which are carcinogenic.[21] The Environmental Protection Agency has listed six carcinogens as having been identified in treated drinking water from various communities.[22] Other chemicals suspected as being carcinogens have also been found in drinking water; more than 70 percent of the 156 other organics found in drinking water have not yet been tested for carcinogenicity—and the full extent of the organic contamination of drinking water supplies in the United States is not yet known. The Hazardous Materials Advisory Committee of the EPA's Science Advisory Board notes that as of mid-1975 "the chemicals so far identified in drinking water account for only a small fraction of the total organic content. Thus, the possibility exists that there may be additional substances in drinking water of equal or greater toxicological significance." The Board concluded that some human health risk does exist from exposure through drinking water. Because of limited epidemiological data and problems with the design of available studies, however, the level of risk could not be quantified.[23]

Some of the organic chemicals already considered in existing water standards because of their acute toxicity—arsenic, asbestos, beryllium, cadmium, chromium, nickel, selenium, and nitrates—are among the substances currently being reviewed for possible carcinogenic effects.[24] Radionuclides are recognized carcinogens, and the EPA estimates that there are no harmless dose levels. Though current standards set no level for sodium, it may be of importance to water consumers who have "confirmed or incipient heart disease, hypertension, renal disease or cirrhosis of the liver."[25] Nor do existing water standards contain specific standards for assessing the health hazards associated with viruses in finished drinking water.[26] Since the analytical procedures for the direct detection of viruses "are not well enough developed nor practicable for widespread application at this time," existing standards for coliforms, turbidity, and free chlorine residuals are relied upon to protect the consumer against whatever hazards may be associated with viruses. Although well-operated water systems applying such standards to the finished product have almost eradicated the non-food-borne enteric bacterial diseases, there has lately been an increasing awareness of water-borne enteric viral disease.[27]

The total number of human enteric virus types which can be water borne is around 100. Water-supply systems containing very low concentrations of virus particles could result in widespread sporadic cases of disease in a community. Though epidemics of some viral diseases are known to have been caused by high concentrations of viruses, the problem of low-level transmission of viruses by water may also be associated with much of the chronic ill health in the United States. A partial list of suspected effects should include acute interstitial myocardites, congenital heart anomalies, and diabetes, as well as the better-known acute diseases thought to be associated with viral agents: infectious hepatitis and viral gastroenteritis.[28] In addition, the role of water-borne viruses in oncogenesis cannot be ruled out.[29]

Technical problems associated with the concentration and cultivation of viruses result in a scarcity of reliable information concerning the efficiency of water treatments as they apply to the removal or inactivation of viruses. What information is available suggests that effective methods for assuring virus inactivation are still in the developmental stage.[30]

For these and other reasons, the EPA has developed a policy (above, Chapter 1) that is opposed to direct recycling for drinking water at this time. Since it is not unreasonable to assume that direct re-use of wastewater is only a difference in degree from what is now practiced when major rivers are used as a raw water source, the EPA policy may also apply to some current and future indirect re-use schemes. L. J. McCabe, Chief, Criteria Development Branch of the EPA, notes:

> All drinking water standards developed to date have cautioned that the limits were set in consideration of using the best source of raw water and cannot be used for guidance for waste water reuse. Thus, any utility that proposes waste water reuse for drinking water in the immediate future will be required to demonstrate the safety of the processes envisioned without a consensus of regulatory authorities on what will constitute a potable water.
>
> A utility gives an implied warranty of the safety of the drinking water they produce and when there are departures from the current state of the art, there will not be much of a defense to fall back on if consumers claim damage from using the water. To provide an adequate defense would require a tremendous amount of analytical work on the water produced. Data would have to be available on any biological or chemical contaminant that some one could think of. The minimal list provided in the drinking water standards would not be enough. It is likely that the analytical cost will exceed the treatment cost. If damages were sought by a consumer, expert witnesses would be required on any conceivable illness of man that could be envisioned to be caused or aggravated by anything that could be measured in water.
>
> Drinking water produced directly from wastewater could be considered to be a new product with many new food additives.[31]

If the effectiveness of any wastewater renovation system is to be evaluated fully before use in a potable supply, complete and proper toxicological and epidemiological evaluations of the finished product will have to be conducted.

28. Liu, p. 7.

29. Ibid.

30. McDermott, pp. 71–73, 76–78.

31. McCabe.

**John T. Reynolds**

32. World Health Organization, pp. 25–26.

33. Ibid., p. 26.

## Toxicological Evaluation[32]

The approach developed by toxicologists in setting tolerance limits for hazardous substances in food has been to establish acceptable daily intake levels for man. These levels are based on all relevant toxicological data available at the time of evaluation. If possible, a "no-effect level" is first determined, and, after adding a safety factor, maximal allowable concentrations are set, taking into account known consumption patterns and realistic levels of contamination from other environmental sources.

Although toxicological methods are available for establishing tolerance levels for contaminants in drinking water, to date little or no use has been made of them. If and when they are used, such toxicological studies will be suspect if they take into account only the acute or subacute effects of specific agents. Evaluations must include the effects of long-term exposure, carcinogenic, mutagenic, teratogenic, and various biochemical and physiological effects of the agent in question and its breakdown products. Other complicating factors are associated with combined and possible synergistic effects of toxic and nontoxic chemicals.

Given the complexities involved in setting tolerance levels for specific contaminants that might appear in a renovated water, such an approach will take a long time to develop and even then might be faulted for not covering all possible effects. A complementary and perhaps more productive approach would involve the toxicological evaluation of the actual product water with its real mixture of residual chemicals. Long-term feeding experiments with a variety of species of experimental animals should be required, as well as other toxicological tests. If such a requirement appears too rigorous, the reader is reminded that most governments require full toxicological evaluations of any new drug or food additive before allowing it to be sold. The requirements for evaluating wastewater intended for re-use involving human consumption with its many unknowns should be at least as rigorous.

## Epidemiological Evaluation

A considerable body of information is available concerning the potential effects on human health of many of the contaminants possibly present when wastewater is used for domestic consumption. But the problems of extrapolating from the possible to the actual remain.

A logical approach to the answers of questions raised on that basis would be to carry out prospective epidemiological studies on populations in water-short areas that were about to go ahead with intentional wastewater re-use for potable supply. Since the opportunities to do such studies will probably be few, retrospective studies may present a more promising alternative, for unintentional and often uncontrolled re-use is and has been widely practiced.

Comprehensive epidemiological studies have been called for, but few, if any, have begun.[33] Such an undertaking is a formidable task. Given the chronic effects possibly associated with the consumption of renovated wastewaters, study design alone presents enormous difficulties.[34] The multifactorial nature of the etiology of chronic disease, with its long latent period, indefinite onset, and the differential effect of a variety of factors—cultural, economic, genetic, physiological, environmental—on the incidence and course of the disease all make incidence data difficult to collect and interpret. Nevertheless, whatever information can be gleaned from such studies would be of great value in providing a basis for the validation of existing standards or developing new ones. Such data however, we must repeat, are not yet available.

## Current Practices in Water Treatment

Very few full-sized advanced treatment plants have been built for continuous routine operation, either to protect existing supplies or to supply re-used water as a new source.[35] Jerome Gavis reports on three plants—in Hanover, Illinois; South Lake Tahoe, California; and Windhoek, Namibia.[36] The varied processes in the plants consistently bring the product water "to a quality equivalent to drinking water except for total dissolved solids and pathogenic organisms."[37] Refractory organics are reduced to about 7 ppm.[38] As has been noted above, there are still unanswered questions as to suitability of such effluents for use in domestic supplies, even though they are clearly better than the raw water now being used in some places.

Part of the answer will involve the availability of practicable analytical procedures for the surveillance required to monitor situations when water-supply and waste-disposal treatments, no matter how sophisticated, must be treated as single systems.[39] Specific questions have been raised as to the availability of suitable methods to detect and quantify viruses and organics for the routine monitoring of water supplies and to permit the evaluation of water and wastewater treatment processes under actual field conditions.[40]

In the case of viruses, "despite the need for considerable further development of existing virus detection methods and the need to run carefully controlled comparative studies on the most promising ones, it can be said that a number of effective methods, which are relatively inexpensive and easy to operate, are currently available for detecting low concentrations of viruses in large volumes of water. Some are being used routinely with good effect for that very purpose."[41] While few water supply systems now have suitable laboratories and personnel,[42] "the possibility does exist that local laboratories can concentrate the sample—freeze the concentrate for later shipment, and then have the assay done at a regional water quality or public health laboratory having complete facilities and staff for virus assay and identification work."[43]

Nevertheless, "the identification of the trace amounts of organic substances remaining in effluents after extensive treatment is beyond the capability of all but the best equipped laboratories, and it is at present not practical or possible to identify all compounds . . . Modern analytical tools such as gas chromatography, mass spectrometry, and nuclear magnetic resonance can

34. For a general discussion, see Judith A. Mausner and Anita K. Bahn, *Epidemiology* (Philadelphia, W. B. Saunders, 1974), pp. 307–335.

35. Gavis, p. 131.

36. Ibid.

37. Ibid, p. 133.

38. Ibid., p. 104.

39. Long and Bell, pp. 220–225.

40. World Health Organization, pp. 38–39.

41. H. I. Shuval and E. Katzenelson, "The Detection of Enteric Viruses in the Water Environment," *Water Pollution Microbiology*, ed. Ralph Mitchell (New York, Wiley-Interscience, 1972), p. 359.

42. World Health Organization, p. 38.

43. Shuval and Katzenelson, p. 359.

44. World Health Or-
ganization, p. 39.

45. George E. Symons,
"That G A O Report,"
*Journal of the American
Water Works Associa-
tion*, 66 (1974), 275–
276.

46. Department of
Health, Education, and
Welfare, *Community
Water Supply Study*.

identify and measure trace substances rapidly, but they are expensive and their use requires considerable skill."[44]

What if drinking-water standards acceptable for re-use situations were to be developed, and practicable and appropriate analytical procedures were available for routine monitoring? The reliability of water-supply operations would still be of paramount importance. If one extrapolates from the 1973 General Accounting Office Report[45] and the EPA *Community Water Supply Study* published in 1970,[46] one has to suspect that there are serious and fairly widespread deficiencies in quality control in many water-supply systems in North America with respect to

—adequate source of supply
—adequate protections of supply sources
—adequate treatment facilities
—adequate trained personnel
—proper operating practices
—quality control of product
—adequate and effective surveillance programs by
  state health departments

It would appear overly optimistic to expect that the demanding technical and managerial requirements for the operation of adequately monitored, fail-safe systems needed to protect the public health when renovated wastewater is used for potable supply will be met without the expenditure of considerable time and money to upgrade existing water supply operations. There is little question that such upgrading can be done, but one has reservations as to whether it will be done.

## Summary

Existing drinking water standards cannot be used as guides for wastewater re-use.

Drinking water standards for re-use situations will have to be developed in the light of epidemiological and toxicological evaluations of the relative import of the low-level chronic effects of the full range of chemical and biological agents that may be present in potable supplies containing renovated wastewater. The comprehensive studies required are just beginning, and there will necessarily be a long wait for useable data.

The lack of appropriate standards and the fact that the sophisticated monitoring systems for viral and organic constituents are still in the developmental stage make it difficult if not impossible to assess completely the efficiency of developing treatment systems.

The overall reliability of existing water-supply operations is questionable.

## Conclusion

Available information, both positive and negative, gives rise to serious uncertainties as to the full range of health hazards possibly associated with re-use for potable supply. Prudence requires that if the public health is to be safeguarded adequately, all other alternatives for the augmentation of potable supplies should be exhausted before such practices are sanctioned.

# Evaluating the Economic Feasibility of Water Re-Use for Urban Water Supply: A Simulation Model*

Daniel Dworkin

**5**

Economic feasibility is clearly a major element in decisions concerning water re-use. Many managers are convinced that once the economic competitiveness of re-use has been demonstrated, adoptions will quickly follow. To date, however, methodologies to evaluate the re-use alternative have been lacking. This chapter presents one means—a stand-by re-use system—of exploiting wastewater renovation and re-use in urban water-supply development. By using a simulation model to compare the fixed schedule of planned expansion of water-supply sources with a system utilizing stand-by re-use, economic feasibility may be gauged. Colorado Springs, Colorado, provides the data basis for the simulation model.

1. L. R. Beard, *Method for Determination of Safe Yield and Compensation Water from Storage Reservoirs*, U.S. Army Corps of Engineers Hydrologic Engineering Center Technical Papers, 3, (Sacramento, California, The Center, 1965), p. 10.

## Rationale for Re-Use

In most cities that re-use their water, the sewage department controls a completely separate water renovation and distribution system. A more flexible alternative is to place control in the water department and provide a method of interchange so that either water from the potable system or renovated effluent could serve nonpotable uses.

Re-use could delay or obviate the need for conventional additions to supply by providing (1) a substitute for high levels of assurance or reliability of supply; (2) a method of mobilizing oversupply resulting from the understatement of system yield; and (3) a method of shortening the planning horizon to allow the pragmatic evaluation of change in demand to replace the present long-term projections.

*Substitution for high assurance levels*. A given stream and reservoir will produce a higher yield as the required level of assurance (the percentage of time when the stated yield may be expected) is relaxed. The resulting increase in yield is a function of the distribution of flows and the level of development of the stream (Table 5.1). To obtain the maximum level of assurance considered necessary for municipal supply, it is necessary to provide storage for at least the most severe drought on record; often extra storage allows for the possibility of experiencing more severe droughts.[1]

*Reprinted by permission of the American Geophysical Union, (*Water Resources Research*, 11 (1975), 607–615.

2. C. S. Russell, D. Arey, and R. W. Kates, *Drought and Water Supply* (Baltimore, The Johns Hopkins Press for Resources for the Future, 1970), pp. 100–101.

3. G. W. Fair, J. C. Geyer, and D. A. Okun, *Water and Wastewater Engineering*, Vol. 1 of *Water Supply and Wastewater Removal* (2 vols., New York, Wiley, 1966), pp. 6–8.

4. R. W. Ockershavsen, "In-Plant Usage Works and Works," *Environmental Science and Technology*, 8 (1974), 420–423.

5. P. M. Berthouex and L. B. Polkowski, "Cost-Effectiveness Analysis of Wastewater Reuses," *Journal of Sanitary Engineering*, ASCE, SA6 (1972), 869–881.

TABLE 5.1

*Storage and Flow Relationships in the Colorado River Basin*

(Flow in billions of gallons daily, storage in millions of acre-feet)
(No deduction for reservoir losses)

| Percent Mean Annual Flow | Resulting Yield | Levels of assurance (percent) | | |
|---|---|---|---|---|
| | | 98 | 95 | 90 |
| | | Required Storage | | |
| 90 | 12.2 | 25.5 | 19.3 | 14.0 |
| 95 | 12.8 | 55.0 | 40.4 | 18.3 |
| 100 | 13.5 | | | 31.2 |

Sources: George O. G. Löf and Clayton H. Hardison, "Storage Requirements for Water in the United States," *Water Resources Research*, 2 (1966), table on p. 340; and previously unreported data in Nathaniel Wollman and Gilbert W. Bonem, *The Outlook for Water* (Baltimore, Johns Hopkins Press for Resources for the Future, 1971), table, p. 256.

As an alternative, a stand-by re-use system could provide this assurance. Water could be supplied from storage until the levels were drawn down so that the remaining capacity and minimum inflows could, with the renovation system, supply the expected demand. When storage levels rose, the renovation of water would be discontinued.

*Evaluation of yield estimates.* In addition to allocating the provision of assurance to the re-use process, the water manager can employ the capacity of the re-use facility to provide a method of appraising yield pragmatically. There has been a long-standing academic debate over the engineering estimates of yield. Social scientists insist that such yields are understated,[2] while engineers pursue more stringent evaluations and cite evidence of systems that have failed at the rated capacity.[3]

Re-use could provide a margin of reserve capacity which would allow for a delay of additions to the system pending a pragmatic evaluation of yield to supplant the engineering estimates. This could result in a substantial delay or could alleviate the need for conventional additions to supply.

*Shortening the planning horizon.* An additional benefit of the delay in increasing system capacity would be the shortened planning horizon in evaluating system demand. This would be possible because the decision to increase the capacity of re-use systems is conducive to much faster implementation than decisions which require diverting additional streams, providing new storage, or developing new ground-water sources. Often rapid chemical treatment can meet requirements for interim re-use capacity.[4] Thus estimates of future needs could be based on more current data, and, because the economies of scale are not as great in re-use as in reservoirs and diversion tunnels, it is possible to increase the capacity as required.[5] This is an increasingly important consideration because of the present low birth rate; indeed, assessments of demand based on the extension of historic growth patterns may well be overestimating future demands for water.

*The economics of an integrated system.* In an integrated system, re-use would serve as a stand-by source, with storage providing water, when available, for the system. This practice would maximize the use of storage with its low operating costs and minimize the production of renovated effluent with its high operating costs. As demand continued to rise, more conventional capacity would be indicated when either the opportunities for the use of effluent were exhausted or the costs of increasing the conventional capacity would be less than continuing to increase the use of effluent. As higher quality effluent was required to meet the demands for environmental quality, there would be increased incentive to re-use water. An integrated system could provide a more rational approach than one in which water from storage and water from re-use were produced and used independently.

## Testing the Assumption: A Simulation Model

To test these hypotheses, a generalized computer model was designed to simulate the water supply and waste treatment of a municipal system (Figure 5.1). Proceeding from the general model, researchers simulated the water-supply system of Colorado Springs. They generated streamflows and rainfall by using the streamflow simulation program HEC-4 provided by the Hydrologic Engineering Center of the U.S. Army Corps of Engineers. Water demand models were based on multiple regressions of twelve years of monthly demand for five sectors of use—residential, commercial, industrial, military, and municipal. The municipal use category included metered municipal use as well as all unaccounted-for variation between the metered amount released from the reservoirs and the amount billed to users. Table 5.2 outlines the regression models, the correlation coefficients, and the levels of significance.

A preliminary set of regressions employed monthly rainfall as one of the independent variables. The significance, as measured by the t test, was low (.02). An index of rainfall, computed monthly as the accumulated departure from mean rainfall, served to include the effect of soil moisture carry-over. The methodology is as follows:

$$RI_i = \sum_{j=1} (R_j - \overline{R_j}) \text{ for } i = 1, 2, \ldots, 12.$$

Where:   $i$ = the current month
$RI_i$ = the value of the rain index in the $i^{th}$ month
$j$ = a summation subscript
$R_j$ = rainfall in the $j^{th}$ month
$\overline{R_j}$ = average rainfall for the $j^{th}$ month

The regression uses a dummy variable season. During the period from May to September, the variable equals 1 when it would be included in the computations. At all other times, season is set at 0 and is not part of the regression.

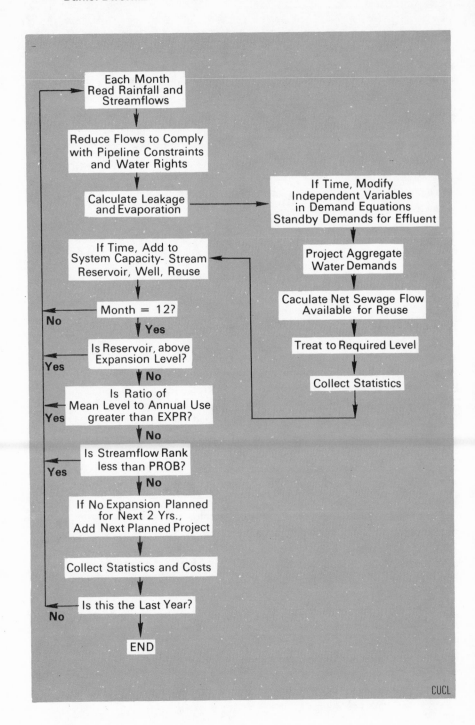

Figure 5.1: Flow Diagram of the Model.

TABLE 5.2

*Summary of Regressions of Monthly Water Use*

| Sector | Percentage of total use (1970) | Correlation Coefficient | Constant | Variables | | | | |
|---|---|---|---|---|---|---|---|---|
| | | | | Time [a] (Monthly) | Season [b] | Price | Population [c] | Index of Rainfall |
| Residential | 40 | .76 | −3.050 | .005 | 7.28 | −.007 | 79. | −.06 |
| Significance [d] | | | .050 | .050 | .05 | .10 | .05 | .0005 |
| Commercial | 17 | .29 | −4.860 | −.041 | 1.048 | −.058 | 1.45 | |
| Significance | | | .005 | .0025 | .0025 | .01 | .0005 | |
| Industrial | 4 | .50 | 1.090 | .007 | .52 | −.005 | 200. | |
| Significance | | | .0005 | .0005 | .0005 | .05 | .0005 | |
| Military | 16 | .67 | .284 | .02 | 2.26 | | | |
| Significance | | | .200 | .0005 | .0005 | | | |
| Municipal | 23 | .22 | .9150 | .24 | | | | |
| Significance | | | .100 | .0005 | | | | |

n = 144
[a]Time is in accumulated months.
[b]Season equals 1 from May to September, 0 all other times .
[c]Industrial use population employment $\times 10^{-4}$.
[d]Significance as measured by $t$ test.

Equations of water demand are based on these models. The unexplained variation is included by adding the mean of the residuals and the product of the standard deviations multiplied by a random normal device.

*Colorado Springs*. The water-supply system of Colorado Springs is a complex network utilizing (1) ground water; (2) surface water from local streams, a Bureau of Reclamation project, and interbasin transfers; and (3) renovated water from returned sewage. Twenty reservoirs, some for the nonpotable system, provide for the storage and release of water. The doctrine of prior appropriations governs the amount and timing of diversions from streams and the storage or re-use of water.

The areas served by the water system have been growing rapidly both by natural population increases and by annexation of the surrounding water systems. In the decade 1960–70 the population of the SMSA which includes Colorado Springs increased 64 percent. To meet the demands, the water system has expanded rapidly from less than 10,000 acre-feet in 1960 to over 50,000 acre-feet in 1973, including the capacity to provide 3600 acre-feet of renovated effluent. Secondary treated effluent is further treated either by filtration to produce irrigation water or by coagulation, sedimentation, and carbon filtration to produce water for industrial use. Present plans are for additions that will bring the total yield from all sources up to 97,700 acre-feet (Table 5.3 and Figure 5.2). This plan for increasing the capacity at fixed times was simulated.

TABLE 5.3

*Future Additions to the Water Supply System*

| Year | Plan | Yield (acre feet) |
|------|------|------------------:|
| 1977 | Eagle-Arkansas | 5,000 |
| 1979 | Homestake (2nd phase) | 17,000 |
| 1985 | Fryingpan-Arkansas (2nd allotment) | 10,000 |
|  | Total | 32,000 |
| 1979 | Re-use | 4,000 |
| 1983 | Re-use | 5,500 |
|  | Total | 9,500 |
|  | Grand Total | 41,500 |
|  | Present capacity | 56,200 |
|  | Present and future Total | 97,700 |

Source: Colorado Springs, Colorado, Water Division records.

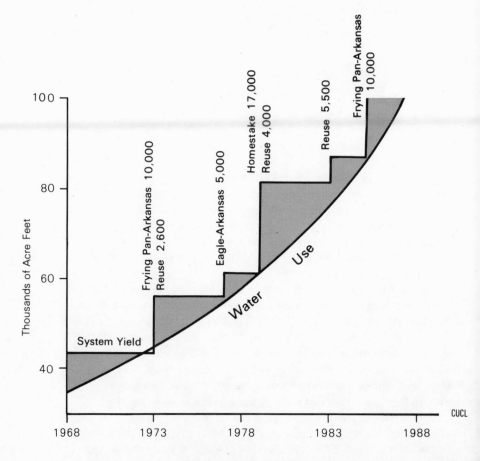

Figure 5.2: Colorado Springs Water Use and Supply Projection.

TABLE 5.4

*Costs of Advanced Waste Treatment*
(Q = millions of gallons of flow daily)

| Process | Construction Costs (Millions of Dollars) | Treatment Costs (Thousands of Dollars) |
|---|---|---|
| Coagulation and sedimentation | costs = .750$Q^{.9004}$ | cost = 44/$Q^{.01746}$ |
| Granulated carbon absorption | costs = .560$Q^{.62487}$ | cost = 136/$Q^{.19438}$ |

## The Alternative

An alternative to the above plan was also simulated. The alternative provided no re-use capacity at the beginning of the simulation and made the
addition of the Eagle-Arkansas and Homestake projects contingent on a decision process based on monitoring reservoir level, streamflow, and water
use. Both the plan and the alternative restricted renovated water to nonpotable uses.

*Decision for re-use: The alternative.* In the model, re-use capacity, if insufficient, is increased, and renovated effluent produced, when the reservoir level
drops below a selected level. When the reservoir level exceeds a set level,
re-use is discontinued. In the period before re-use capacity is added and
when re-use is subsequently discontinued, potable water is furnished to
all nonpotable users. When reservoir levels are low, renovated nonpotable
effluent partially meets the total monthly demand for each sector. Costs of
increased capacity of the re-use system are invested in the first year of any
decade and are based on the peak daily flow during the decade. Construction and operational costs are calculated separately (Table 5.4).

*The decision process for the Eagle-Arkansas and Homestake projects.* At
the end of each year in the simulation, a decision is made on whether or not
to add to the potable supply system. Three checks are made: is the reservoir
level higher than required for expansion, is the amount of water in storage
during the year greater than a set percentage of the annual use for the year,
and is the total streamflow for the year less than a percentage of the combined record of historical and synthetic flows to the present? If any of the
comparisons are positive (that is, if reservoir levels are higher than the set
expansion level, if the ratio of water level to total use indicates a higher
percentage of annual use in storage, or if the rank of flows is less than might
be expected during a set percentage of the time), there will be no expansion.
If all are negative, a further check is made for the status of the second
Fryingpan-Arkansas allotment. If it is scheduled for addition to the system
within the next two years, neither of the other two plans will be used. If it is

not, then the Eagle-Arkansas or, if this has already been added, the Home-stake project will be scheduled for addition after two years. The two-year delay had been selected for Colorado Springs because much of the engineering and most of the diversion tunnels had been completed as part of other projects. To test the results researchers ran an additional series of simulations using a four-year delay.

## The Simulation Results

Both the plan and the alternative were simulated for three streamflow series differing from one another in the frequency, timing, and duration of droughts. Series 1 provides a low-flow period at the end of a 50-year simulation, when demand for water will be greatest. Series 2 has a prolonged low-flow period during the mid-years of the 50-year period. Series 3 represents a low mean flow for all except the first 15 years. The statistics for the three streamflows appear in Table 5.5.

Three population series were used. The high population projection provided the population growth predicted by the consultants to the water division. Using the water demand models developed for the simulation and the high population projection, the resulting water demands were higher than the projections of the consultants who are projecting a lowered per-capita use not yet apparent in the historical data (Figure 5.3).

The medium population projection provides the estimated water use. A third projection assumes a lower growth for the system. The Pike's Peak

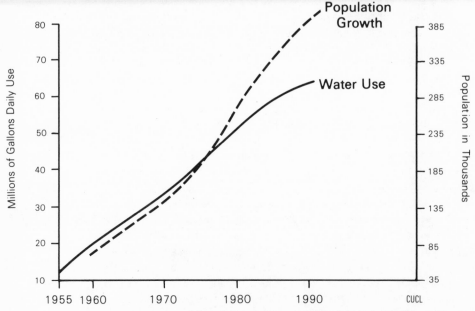

Figure 5.3: Population Growth and Daily Water Use. Projections by Black and Veatch. Graphic estimating curve by the author.

TABLE 5.5

*Streamflow Statistics*
(all flows in cubic feet per second)

| Series | Mean Flow | Standard Deviation | Maximum Flow | Minimum Flow |
|---|---|---|---|---|
| Series 1 | 3498 | 1343 | 6760 | 2100 |
| Series 2 | 3475 | 1266 | 6625 | 2100 |
| Series 3 | 3272 | 1055 | 6471 | 2100 |

Area Council of Governments has prepared high, medium, and low projections for El Paso County to the year 2000. These were compared with the
populations simulated in the model. A comparison between the high and
medium populations indicates that the water system is projected to serve
approximately 70 percent of the population of the county projected by the
Pike's Peak Area Council. The low growth series assumes that the water
system will be serving nearly the entire projected population of the county by
the year 2000 (Table 5.6).

These simulations made no distinction between the value of water from
storage and water from the nonpotable system. The comparison between
the plans was based on the present value of costs, both capital and operating, to maintain the supply system for the period of the simulation.

## The Results

*The high population projection.* Under the assumption of high population
growth, the plan could not support the demands over a period of 50 years.
Using three streamflows, the plan failed after 33, 34, and 36 years. The
alternative, however, supported demands for a full 50 years.

TABLE 5.6

*Comparison of Population*
Pike's Peak Area Council of Governments for El Paso County and
population projected in Simulation Model
(population in 1000's at year's end)

| Year | High | | | Medium | | | Low | | |
|---|---|---|---|---|---|---|---|---|---|
| | Council | Model | % | Council | Model | % | Council | Model | % |
| 1975 | 312 | 211 | 68 | 310 | 206 | 66 | 293 | 206 | 70 |
| 1980 | 381 | 277 | 73 | 368 | 250 | 68 | 314 | 246 | 78 |
| 1985 | 467 | 333 | 71 | 417 | 286 | 69 | 328 | 277 | 84 |
| 1990 | 562 | 397 | 71 | 459 | 319 | 69 | 343 | 307 | 90 |
| 1995 | 673 | 450 | 67 | 504 | 343 | 68 | 354 | 334 | 94 |
| 2000 | 797 | 489 | 61 | 551 | 366 | 66 | 365 | 355 | 97 |

6. Black and Veatch
Consulting Engineers,
*Report on Cheyenne
Canyon Booster District
and Templeton Service
Area for Colorado
Springs, Colorado*
(Kansas City, Missouri,
1972), p. 15.

The alternative delayed the additions of the Eagle-Arkansas and Home-stake projects an average of 16 years. The capacity of the re-use plant was nearly three times larger (than required for operating the plan), but was not required until later. Only in the fourth decade did the amount of water re-used exceed that scheduled for the plan. The present value of the costs of imple-menting the plan averaged $21 million, the alternative accumulated costs averaging $12 million (Table 5.7).

*The medium population projection.* The medium population projection re-sulted in a level of water use close to that predicted by the engineering con-sultants to the city for the year 1990.[6] The system was able to support the water demands projected by the simulation for the complete 50-year run. The costs, since they represent scheduled additions to capacity which are independent of streamflows or demands (except when demand causes the system to run out of water), were the same for all the series: $21.5 million. The different streamflow series used in the simulation affected the alterna-tive, in contrast to the plan. Both the timing and capacity of the plant for re-use, as well as the timing of the additions of the Eagle-Arkansas and Homestake projects, were affected (Figure 5.4). The present value of the costs incurred were nearly equal, but the components and timing varied sig-nificantly.

*The low population projection.* The plan again required $21.5 million, whereas the alternative required an average investment of $6.3 million. Using the low population series, the Eagle-Arkansas project was delayed for an average of 33 years, and the Homestake project was never required.

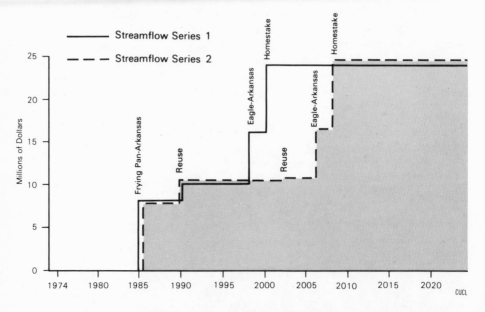

Figure 5.4: Comparative Timing and Amount of Investment, Medium Population; Streamflow Series 1 and 2.

TABLE 5.7

*Summary of Simulations: High, Medium, and Low;*
*Population Assumptions: Streamflow Series I*

| | The Plan[a] | The Alternative | | |
| --- | --- | --- | --- | --- |
| | | High Population | Medium Population | Low Population |
| *Years of Projection* | 50 | 36 | 50 | 50 |
| *Capacity of Re-use Plant (Maximum)* | | | | |
| *Millions of Gallons/Day* | | | | |
| Decade 1 | 12 | 0 | 0 | 0 |
| 2 | 12 | 26.4 | 0 | 0 |
| 3 | 12 | 27.4 | 11.9 | 9.7 |
| 4 | 12 | 32.0 | 11.9 | 12.8 |
| 5 | 12 | | 11.9 | 16.8 |
| *Re-use Water Processed* | | | | |
| *Millions of Gallons* | | | | |
| Decade 1 | 19,000 | 0 | 0 | 0 |
| 2 | 40,000 | 7,000 | 0 | 0 |
| 3 | 40,000 | 10,644 | 1,400 | 700 |
| 4 | 40,000 | 40,107 | 0 | 1,500 |
| 5 | 40,000 | | 0 | 2,100 |
| *Additions to Capacity (Year)* | | | | |
| Fryingpan-Arkansas | 1985 | 1985 | 1985 | 1985 |
| Eagle-Arkansas | 1977 | 1993 | 1998 | 2013 |
| Homestake | 1979 | 1998 | 2000 | never required |
| Present value of investments (millions of dollars at 6.875 per cent) | | | | |
| Re-use Capacity | 2.20 | 1.91 | 0.7 | 0.6 |
| Re-Use Operations and Maintenance | 4.44 | 1.69 | 0.4 | 0.8 |
| Conventional Capacity | 14.50 | 7.17 | 6.5 | 4.6 |
| | 21.50 | 10.77 | 7.6 | 6.0 |

[a]These figures are for 50 years of operation of the plan. There is little difference, less than a million dollars, in operating the cost of the plan for less than fifty years.

*The 48-month delay.* Using Streamflow Series 1, a 48-month delay before increasing conventional capacity was simulated. This increased the present value of the operation and the maintenance and capacity costs of the re-use system, partially offset by the delay and the resulting decreased present value of the conventional capacity. Since this affected the latter years of the simulation, the changes in the present values were small, an increase of $.2 million for the high population projection and less for the medium and low population series.

## Summary

For every assumed population and streamflow condition simulated, the alternative offered a less costly method of supply than the plan. The costs of the alternative varied according to levels of supply and demand, but the costs of the plan were independent of these conditions.

Of the two methods of providing water for Colorado Springs, the plan is less desirable under conditions of low population growth. Under the assumption of high population growth, the plan failed in the fourth decade of operation. The alternative could supply demands for the full 50 years at a cost of $12 million or less (Table 5.8).

TABLE 5.8

*Summary of the Six Simulation Series:*
*the Plan and the Alternative*
(millions of dollars at 6.875 percent)

| Streamflow Series | Population Projection | Year of Simulation | Present Value | |
|---|---|---|---|---|
| | | | The Plan | The Alternative |
| 1 | High | 35 | 21.1 | 10.8 |
| 2 | High | 32 | 20.9 | 10.0 |
| 3 | High | 33 | 21.0 | 12.0 |
| 1 | Medium | 50 | 21.5 | 7.6 |
| 2 | Medium | 50 | 21.5 | 7.5 |
| 3 | Medium | 50 | 21.5 | 8.1 |
| 1 | Low | 50 | 21.5 | 6.0 |
| 2 | Low | 50 | 21.5 | 5.9 |
| 3 | Low | 50 | 21.5 | 6.9 |

The alternative is economically a more efficient method of providing a water supply for Colorado Springs. The additional expense of the plan arises from the cost of treating effluent for re-use when water from the potable supply system is available in sufficient quantities to meet the demands of all users and from the investment in conventional and re-use facilities, which could have been delayed. The alternative plan employs water processing only when the supply in storage is low and delays additions to conventional capacity until parameters measuring the supply and use of water indicate such a need. To provide further protection against premature investment, the alternative precludes any decision for expansion during the extreme low-flow years.

This chapter has presented a method of integrating the renovation and re-use of effluent into the water system and has compared such a system with the planned system in Colorado Springs. The integrated system emerges as a more efficient method of planning than the present system because (1) the former would allow the demands on the system to rise without increasing conventional capacity, thus using any reserve capacity in the system; and (2) it could substitute for high levels of assurance.

The simulations indicate that the integrated system is, in fact, far more efficient. This suggests that a minimum requirement of present planning for the renovation or re-use of water is the modification of systems to include a physical cross-connection between the two, allowing water to be used interchangeably for nonpotable purposes; and a common management process to make such decisions.

# Wastewater Reclamation and The Water Economy in Israel: A Case Study [1]

**6**

Stephen Feldman

Thoughts of wastewater re-use have been in the minds of Israeli engineers for nearly thirty years since the establishment of the state itself in 1948. Compared with today's standards, Israel at the time was utilizing a relatively small proportion of her total water potential, yet Israeli planners foresaw the future limitations imposed on consumption by the diverse but at best semi-arid geography within the bounds (7,000 square miles) of the state. With the onset of the 1960's Israel found itself by necessity undertaking research aimed at producing unconventional water supplies to compensate for the nation's limited physical resource base.

The period between 1948 and 1975 has witnessed a profound demographic growth in the state. Net immigration of some 1,300,000 persons ballooned the total population from 873,000 in 1948 to 3,300,000 in 1975. This immigration has been accompanied by a relatively high rate of natural population increase (Jews and non-Jews)—21.6 per 1000 as compared to 9.9, 8.0, 7.2, and 6.4 for Canada, the United States, Italy, and France, respectively.[2]

These population pressures impose drastic burdens upon the nation's water economy. A sharp change in temperature and precipitation over relatively short geographic distances characterizes Israel's water budget. Climatically the nation experiences a large drop in precipitation from north to south. Since precipitation prevails from November to April, it is thus inversely related to temperature. Forty percent of total water consumption occurs in the relatively water-rich northern areas, which contain about 60 percent of the nation's water supplies. The southern regions have substantial deficits in production but contain most of the irrigable land, upon which rests most of Israel's potential for export crops, particularly citrus. The regional deficits of the south are offset primarily by water pumped through the National Water Carrier, the focal point of Israel's highly integrated attenuated supply system. Figure 6.1 shows the geographic extent of the Carrier.

The dependence of these irrigated areas upon that supply is illustrated by the increases in water pumped through the Carrier since its inception in 1964. The 1971/72 withdrawals are shown in Table 6.1. Table 6.2 shows the regional distribution of water production corresponding to Figure 6.1. The pumping of water is carried out primarily during the dry summer months and artificial recharge of ground-water aquifers occurs during the months of off-peak usage. Up to 30 percent of the Carrier's capacity, or about 8 percent of total national withdrawals, undergoes artificial recharge. The role of recharge

1. The author is indebted to Mr. Peretz Darr for supplying necessary unpublished materials, Gil and Hella Yaneev for aid in data collection, and a grant from the Hebrew University Social Science Council for partial funding of this research.

2. United Nations, *Statistical Yearbook* (1972).

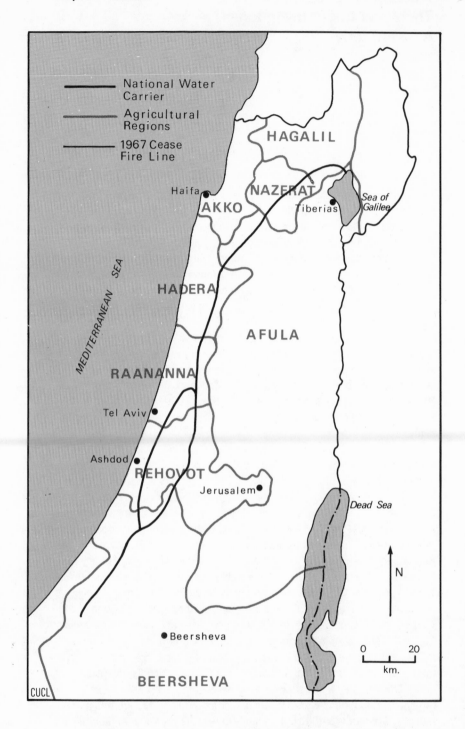

Figure 6.1: The National Water Carrier.

TABLE 6.1

*Water Consumption and Production by*
*Agricultural Region, 1971/72*
(millions of cubic meters)

| Region | Production[a] | Consumption[b] |
|--------|------------|---------------|
| Beersheva | 108 ( 6.2) | 265 (16.9) |
| Rehovot | 222 (12.7) | 258 (16.5) |
| Raanana | 280 (16.1) | 286 (18.3) |
| Hadera | 206 (11.9) | 218 (13.9) |
| Afula | 660 (38.0) | 301 (19.2) |
| Nazerath | 2 ( 0.1) | 9 ( 0.6) |
| Akko | 110 ( 6.3) | 76 ( 4.9) |
| HaGail | 151 ( 8.7) | 152 ( 9.7) |
| Total for country | 1740 (100.0) | 1565 (100.0) |

[a]Parentheses contain percentages of total for the country.
[b]Inclusive of 149 million cubic meters of recharged water.
Source: Israel, Ministry of Agriculture, *Water in Israel*,
Part B (Tel Aviv, 1973), p. 120.

juxtaposed against re-used wastewater will increase in the near future; the
discussion of this relationship will be brought out below.

The main pumping station of the Carrier (at its widest some 108 inches in
diameter) lies approximately 210 meters below sea level and pumps water
from there to a height of 150 meters above sea level. This method of water
transfer to the nation's southern regions has proven to be the cheapest
means of supply and represents the final stage in the conventional development of the national water system.

Given one of the most rapidly expanding economies in the world and a
population expected to reach some five million by 1990–95, projections of

TABLE 6.2

*Production of the National Water Carrier, 1964/65*
*to 1971/72*
(millions of cubic meters)

| Year | Production |
|------|-----------|
| 1964/65 | 139 |
| 1965/66 | 269 |
| 1966/67 | 283 |
| 1967/68 | 332 |
| 1968/69 | 320 |
| 1969/70 | 353 |
| 1970/71 | 380 |
| 1971/72 | 397 |

Source: *Water in Israel*, Part B, p. 146.

3. E. Kally, "Israel's Water Economy and Its Problems in the Early Seventies," and J. Bonne and Z. Grinwald, "The Estimated Growth of Water Consumption in Israel," in *Water in Israel*, Israel, Ministry of Agriculture (Tel Aviv, 1973), Part A.

4. See, for example, H. Zvi Griliches, "Research Expenditure, Education and the Aggregate Agricultural Production Function," *American Economic Review*, 54 (1964), 961–974.

5. Much of this research in Israel has been conducted by the Faculty of Agriculture, The Hebrew University, and the Mekorot Water Company's research division.

6. The term "demands" is not defined in the strict economic sense, but, as in Chapter 1, its usage here is more synonymous with "requirements." See P. Darr, S. Feldman, and C. Kamen, *Operational Approaches to Urban Water Supply: A Regional Analysis of Water Supply Capacity Design in Israel* (Rotterdam, Holland, Rotterdam University Press, 1976).

water withdrawals ring a somber note for Israeli water resource managers. According to Table 6.3, the sectors of the water economy growing at the most rapid rate are the urban (domestic, commercial, and public) and industrial.[3] Domestic consumption has been growing at a rate of 4 percent annually. Grossly analyzed, half the growth has resulted from population increase and half from rising standards of living. Consumption increases at the fastest rate in industry, where 10 percent per annum is normal. Industrial output itself is growing at 12 percent per annum. Stringent metering and the requirement of licenses for industrial withdrawals exceeding 5,000 cubic meters per annum ensure the practice of water conservation within plants. Since water consumed in these two sectors had the highest marginal value in use, curtailment in the agricultural sector seems likely. Agriculture will have to be flexible in adapting to tighter supply conditions. It has been apparent for some time that the assessment of wide varieties of alternatives in the production function for agricultural crops leads to greater efficiency of water use.[4] Experiments on salinity tolerance, crop hybridization, more intensive use of fertilizer, automatic valve controls, and drip and aerosol sprinkling as substitutes for fresh water have shown that less water need be applied per unit of output.[5] Large-scale application of these techniques has been undertaken with great success. As partial evidence of increased water efficiency, Table 6.4 shows the decrease in water consumption per unit of agricultural land. Since 1958 approximately 45 percent of the irrigable acreage has increased, whereas agricultural consumption has increased by only some 20 percent, or less than half that rate. Concurrently, the value of average crop output per cubic meter of water in agriculture has risen from I£0.63 in 1962 to I£1.10 in 1971 (one U.S. dollar equals I£4.20). Water consumption per I£1,000 of crop production dropped from 1,598 cubic meters to 913 cubic meters. These figures represent not only improvement in crop mixes, but also gains in water-using efficiency through the above techniques. Whereas agriculture may be flexible with regard to economic considerations, other associated aspects may be more rigid. Security, ideology, and political considerations provide reasons for supplying water at costs that exceed its marginal value product in use.

Projections of demands[6] in consumption, therefore, foresee deficits in the early 1980's and perhaps sooner if poor rainfall prevails for two or three successive years. Table 6.5 presents projections based upon past trends. Given that consumption will presumably be greater than supply, the range of choice of supply technology needs to be examined.

## The Water Economy: Retrospect and Prospect

As already noted, the sectors of the water economy growing at the most rapid rate are the urban (domestic, commercial, and public) and industrial. Domestic withdrawals have been growing at 4 percent annually. As mentioned above, withdrawals for industry have been increasing nearly as fast as the annual increase in Israel's industrial output: 10 percent for the former and 12 percent for the latter. Juxtaposed against Table 6.2, Table 6.5 may

TABLE 6.3

*Total Water Consumption by Use Category*[a]
(in millions of cubic meters)

| Year | Total | Agriculture | Domestic | Industrial |
|------|-------|-------------|----------|------------|
| 1958 | 1274 | 1032 | 196 | 46 |
| 1971/72 | 1565 (23) | 1210 (17) | 268 (37) | 87 (89) |

[a]Figures in parentheses indicate percentage change since 1958.
Source: *Water in Israel*, Part B, p. 54.

TABLE 6.4

*Irrigated Area and Average Water Consumption of Irrigated
Land on a Per Dunam Basis for 1958 and 1971*[a]

| Year | Irrigated Area (thousands of dunams) | Average Water Consumption Per Dunam (cubic meters) |
|------|-------------------------------------|---------------------------------------------------|
| 1958 | 1185 | 871 |
| 1971 | 1735 | 697 |
| Percentage since 1958 | 46.4 | −20.0 |

[a]One dunam equals one-quarter of an acre.
Source: *Water in Israel*, Part B, p. 26.

TABLE 6.5

*Water Consumption and Production in Israel in 1972
and The Mid-1980's*
(millions of cubic meters)

| Consumption or Production Category | 1972 | Mid-1980's |
|-----------------------------------|------|-----------|
| Domestic | 270 | 470 |
| Industrial | 70 | 230 |
| Total urban (domestic plus industrial) | 340 | 700 |
| Yield of sources[a] | 1450 | 1700 |
| Left for agriculture in water balance | 1110 | 1000 |
| Actual agriculture consumption | 1180 | — |
| Sewage effluents outside Dan Region available for use[b] | 30 | 250 |

[a]Prospective yield of replenishable water resources only, without
surpluses from previous rainy years and without "one-time" reserve.
This yield is conditional upon the development of new resources,
including reclaimed effluents of the Dan Region sewerage system and
utilizations of artificial rainfall in the Lake Kinnerot basin on a very limited
basis.
[b]Actual consumption of sewage effluents.

7.  G. E. Levite, ed., *Developments in Desalination Technology in Israel* (Jerusalem, National Council of Research and Development, 1971).

8.  Stephen Feldman, "Artificial Rainfall Stimulation: A Comparative Systems Analysis of Israel and the U.S. Southwest" (unpublished).

9.  From A. Feinmesser *Survey of Sewage Collection, Treatment and Utilization* (Tel Aviv Water Commission, Ministry of Agriculture, 1971).

be compared as the straight-line projections are made for the mid-1980's by source of supply. These figures can further be compared with the potential of water resources based on available hydrological data (some 35 years for river flows, 20 years for ground water, and 15 years for small streams). The average annual water potential is estimated at 1,500 million cubic meters. Given the variance associated with these data between wet and dry years, nearly 3,500 million cubic meters are needed for long-term regulations. Indications are that all or most of the 1,500 million cubic meters are currently being exploited. Table 6.5 indicates the range of choice in expanding sources of supply. The most popular are:

1.  re-used wastewater
2.  artificial rainfall
3.  brackish waters
4.  marginal supplies, including deep-bore wells and storm-water runoff catchment
5.  desalinated seawater
6.  the one-time "mining" of stock water resources from aquifers.

With inland movement of the salt water–fresh water interface, the last, most undesirable, alternative has been occurring to some extent. Numerous desalination facilities do exist in Israel, but they are of small scale and are confined to settlements where fresh water is lacking. The largest plants thus far are the installations for the Red Sea port of Eilat.[7] It would appear that the long-run solution in this technology lies in the use of relatively inexpensive atomic power. Since financial and technical plans are only at the nascent stages of development, however, it seems improbable that any sizable installation will be constructed by 1985–90.

Artificial rainfall appears to be a partial solution for even immediate application. Feldman has reviewed the prospects for this technology and reports its low cost,[8] and Israel's favorable cloud physics environment makes this a most appealing alternative. The resolution of the external effects (discussed in Chapter 1) upon neighboring states from rainfall stimulation does need, however, much more research than currently exists. The projections of Table 6.5 do include modest abetment from such activity, which is under way over the Kinnerot basin.

The technology that is commanding the most financial resources within Israel's water economy is wastewater re-use.

*Wastewater re-use in Israel: The scope of activity.* In 1971 Israel undertook a nationwide survey to determine the water potential of the existing sewage flows, the percentage of flow already utilized, and the determination of future possibilities for their utilization. The survey falls into two principal sections: (1) a survey of urban sewage and sewerage, and (2) a survey of sewage in the rural sector.

The survey evaluated 79 urban settlements with a combined population of approximately 2.5 million and an estimated effluent discharge of 301,000 cubic meters per day, or about 110 million cubic meters per annum.[9] Public sewers convey approximately 78 percent of the total, a volume equal to total urban effluent excluding the industrial sector. Approximately 34 percent of the wastes undergo primary treatment. Almost all of the existing facilities are underdesigned to meet their target populations. Currently, about 134,000 cubic meters per day of sewage, 29 percent of the potential in urban areas, is reclaimed. Some 5 percent of this is used for ground-water recharge and 31 percent for agriculture. These figures are exclusive of the Dan Region Water Reclamation Project, which will be discussed separately below.

Table 6.6 shows the amounts of sewage collected, treated, and utilized in the urban sector for 1963, 1967, and 1971. The remarkable increase in sewage from 1963 to 1971 reflects the general increase in water use, as reported above. It should be pointed out, however, that the earlier survey does not take into account industrial wastes not disposed of through urban systems. This tends to bias those figures downward. Of the 79 urban settlements surveyed, some 77 have partial or complete central sewage systems, 75 have at least some level of treatment, and 46 towns reclaim some of the effluents. Tables 6.7 and 6.8 illustrate water re-use for irrigation and recharge purposes for the country as a whole in the towns surveyed inclusive of the Dan Region.

The national survey covered 226 kibbutzim, 33 moshavim, and 20 collective moshavim, each representing forms of communal agriculture, as well as 47 agricultural schools and other institutions which have central sewerage systems. Wastewater was estimated on the basis of the number of inhabitants, number of livestock and poultry, nature of industry, water consumption, and various other conditions. Table 6.9 shows the amounts of sewage in the rural sector, 316 settlements containing some 116,000 inhabitants plus another 10 percent nonresident population. Out of the total of 316 settlements, 149 re-use their wastewater and six more plan to do so in the near future.

As of 1971 the total quantity of re-used wastewater by both the urban and rural sectors totaled 166,000 cubic meters per day, some 32 percent of the potential amount. On an annual basis the latter figure is about 20 percent of the total amount, the difference between the daily and annual figures being due to the fact that in winter the sewage is rarely used for irrigation and flows into outfalls. The major treatment method utilized in all parts of the country, especially in the small-to-medium-sized settlements, is the oxidation pond. These ponds, aerobic and usually linked to anaerobic ponds, are relatively inexpensive outside of the major urban areas, and they double as storage reservoirs for the irrigation water. Already some 220 such ponds are in operation, a 500 percent increase from the previous decade. There is one major drawback. Land area in the coastal plain near the major population centers is indeed scarce and costly, and the increase in land use for this purpose may bring about higher relative costs for this particular method of disposal in certain areas.

*The Dan Region Water Reclamation Project*. The most important ongoing (operating) sewage treatment project in the country is the Dan Region Water Reclamation Project. The project covers seven towns forming the Greater Tel Aviv Metropolitan Area, with an anticipated 1990 population of greater than one million. Ultimately the project will handle the disposal of well over

**Stephen Feldman**

TABLE 6.6

*Amounts of Sewage in the Urban Sector*

| Description | 1963 cu.m/day | % | 1967 cu.m/day | % | 1971 cu.m/day | % |
|---|---|---|---|---|---|---|
| Total | 339,550 | 100 | 343,170 | 100 | 451,805 | 100 |
| Sewered | 240,045 | 70 | 287,540 | 84 | 352,185 | 78 |
| Treated | 88,440 | 26 | 119,080 | 35 | 155,300 | 35 |
| Total utilized: | 48,470 | 14 | 79,670 | 23 | 133,535 | 29 |
| a. In agriculture | 38,070 | 11 | 54,820 | 16 | 95,685 | 21 |
| b. In recharge | 10,400 | 3 | 24,850 | 7 | 37,850 | 8 |

Source: A. Feinmesser, *Survey of Sewage Collection, Treatment and Utilization* (Tel Aviv, Water Commission, Ministry of Agriculture, 1971).

TABLE 6.7

*Agricultural Utilization of Urban Effluent by Crops*

| Crop | Irrigated Area Dunam | % | Effluent Utilized cm.m/day | % |
|---|---|---|---|---|
| Field crops | 15,330 | 61.5 | 45,680 | 47.7 |
| Orchards and vineyards | 1,400 | 5.6 | 4,900 | 5.2 |
| Citrus | 4,315 | 17.2 | 19,500 | 20.6 |
| Various | 1,520 | 6.1 | 5,150 | 5.5 |
| Sown pasture | 1,330 | 5.3 | 10,075 | 10.0 |
| Fodder | 1,070 | 4.3 | 9,130 | 9.7 |
| Fish ponds | — | — | 1,250 | 1.3 |
| Total | 24,925 | 100.0 | 95,685 | 100.0 |

Source: *Survey of Sewage Collection, Treatment and Utilization.*

TABLE 6.8

*Re-Use of Urban Effluent*

| Use | cu.m/day | 1000's cu.m/year[a] |
|---|---|---|
| Agriculture | 95,685 | 17,040 |
| Recharge[b] | 37,850 | 13,630 |

[a]Year means agricultural year.
[b]Effluents recharged into ground water through infiltration ponds planned for the purpose, or through natural depressions in the ground.

Source: *Survey of Sewage Collection, Treatment and Utilization.*

TABLE 6.9

*Amounts of Sewage in the Rural Sector*

| Source | Potential | | | Sewered | | | Utilized | | |
|---|---|---|---|---|---|---|---|---|---|
| | cu.m/d | and | In 1,000 cu.m for season[a] | cu.m/d | % of potential | In 1,000 cu.m for season[a] | cu.m/d | % of potential | In 1,000 cu.m for season[a] |
| Domestic and Industrial | 49,875 | 100 | 10,000 | 43,600 | 87 | 8,700 | 19,190 | 38.5 | 4,450[c] |
| Swimming pools | 14,600 | 100 | 2,200 | 6,060[b] | 42 | 910 | 13,130 | 90 | 1,970 |
| Total | 64,475 | 100 | 12,200 | 49,660 | 77 | 9,610 | 32,320 | 48.5 | 6,420 |

[a]Assuming the season for re-use numbers 200 days.
[b]Swimming pools connected to sewerage systems.
[c]Including utilization for fish ponds during 265 days per annum.
Source: *Survey of Sewage Collection, Treatment and Utilization*.

50 million cubic meters per annum, of which over 90 percent is expected to be reclaimed for re-use. The major parameters of the project in the first phase include trunk sewers ranging from 28 to 72 inches in diameter, some 15 kilometers of concrete pipe, a cutting screen, and a pumping station discharging through an 880-meter steel pipe 60 inches in diameter. The second phase includes 60–80 inch interception ten kilometers long, with three pumping stations, the largest of which is designed for 21,000 cubic meters per hour in a 55-meter lift and a 16-kilometer long 70–80 inch diameter pipeline for conveyance by gravity.

Figure 6.2 shows the actual design of the Dan scheme. Because of pollution of the coastal aquifer by nitrates, treatment of the effluent is designed to remove this hazard. After this treatment the water is recharged to the sandstone aquifer, where it undergoes a natural-cleansing, "tertiary" process. The nature of treatment in this scheme brings water quality to exceptionally high standards:

*The National Plan*. The successful design of the Dan scheme has shown Israeli planners the virtues of regional re-use projects. Regional schemes consider only intraregional parameters and may not account for the economies of scale that may be evident with the integration of the regional schemes and the national water supply. In 1972 Tahal prepared pre-feasibility reports on the utilization of treated wastewater in central and southern Israel.[10] The planning horizon for the design spans thirty years, but the full capacity of the system is designed to be met by 1985. Naturally, ground-water contamination is exempt from planning horizons, since the national policy is to avoid causing irreversible consequences. The pre-feasibility report also assumes that the effluent discharged from the urban treatment plants will have undergone secondary treatment and constitutes the "source" water for the scheme. It is also assumed that the effluent will be suitable for the irrigation

10. These reports are unpublished and consist of internal memoranda (in Hebrew and English) of Tahal, for the supply of which I am indebted to Mr. Peretz Darr.

**Stephen Feldman**

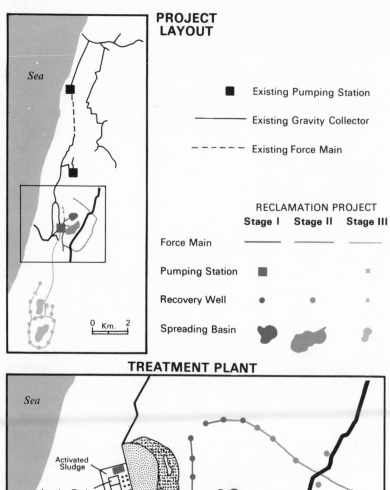

Figure 6.2: The Dan Region Water Reclamation Project.

of such field and industrial crops as animal feed and cotton, which will not come into direct contact with the consumer. Citrus plantations already make considerable use of effluents for irrigating, yet there is no prospect for expansion of such use. There are two primary reasons: (1) the higher salinity of the effluents is likely to cause a significant reduction in yields, and (2) irrigation with effluents may receive unfavorable reception abroad in Israel's export markets.

The National Plan includes three major aspects: (1) utilization of effluents within the region where they are produced, (2) conveyance of effluents from north to south, with the whole use concentrated in the south and with emphasis upon the Negev, and (3) a combined system whereby effluents of the northern regions will service first those regions where they are produced, and surplus effluents will travel south for storage and subsequent utilization in the south.

Nine regions underwent preliminary studies. Table 6.10 shows the results of the analysis for 1985. Hourly flows were computed by dividing the forecast annual effluent flows (1985) by 25 percent. This reflects the winter-summer differential in which minimum demands occur during the winter for the effluent. Local utilization schemes do have the advantage of minimizing conveyance costs for final delivery.

The economics of the regional utilization schemes were examined with reference to five regions: (1) Natanya, including Natanya and Even Yehuda; (2) Southern Sharon, inclusive of Herzliyya, Ramat HaSharon, Kefar Sava, and Hod HaSharon; (3) Ayyalon Association, inclusive of Ramle, Lod, Yehud, Or Yehuda, Kefar Habad; (4) Dan-Ashdod, inclusive of the Dan scheme, Ashdod, and Yavne; and (5) Jerusalem. The analysis appears in Table 6.11. The column headed "Cost of Disposal with No Utilization Scheme" reflects the shadow price of urban sewage disposal. This cost would be incurred anyway to avoid intolerable levels of ground-water and stream contamination. The total summer utilization of 37 million cubic meters has a mean cost of 18 Ag. per cubic meter (100 agorot equals I£1.00). This does not include intrasettlement distribution cost. All of the costs are average costs, whereas marginal costs may be some 30 to 40 percent higher.

Peretz Darr has shown that the absolute difference in unit costs between a single-stage, activated-sludge system and a three-stage, activated-sludge plant with the addition of phosphorous removal, filtration-activated carbon, recharging of the aquifer, and repumping is estimated at 67 Ag. per cubic meter.[11] He questions the transportation distance at which the unit cost of secondary effluent production and long-range transmission equates with those costs associated with the production of potable water. Using a mathematical relationship established by Kally,[12] Darr defines a unit transportation cost per cubic meter per kilometer at .27 Ag with a designed annual flow of 30 million cubic meters.[13] At this transportation cost, given that the break-even point between secondary effluent irrigation and full treatment to potable water is 67 Ag. per cubic meter, the break-even transportation distance is 250 kilometers. The analysis points to a distinct advantage of supplying secondary effluent directly to agriculture or industry, since conveyance distances in Israel are invariably less than 250 kilometers.

11. Peretz Darr [Dalinsky], "Socio-Ecologic Aspects of Closed Wastewater Utilization Systems," *Proceedings of the Fifth Scientific Conference of the Israel Ecologic Society* (Tel Aviv, 1974).

12. E. Kally, "Cost of Conveying Water in Pressure Mains" (unpublished Tahal memorandum, Tel Aviv, 1972).

13. Darr.

**Stephen Feldman**

TABLE 6.10

*Amounts of Urban Effluents in 1985 by Producing Areas*
(in 1,000 cubic meters)

| Region | Town | Production MCM p.a. | Total Production MCM p.a. | Hourly Production (7,000 hrs) cubic meters |
|--------|------|------|------|------|
| Natanya | Natanya | 7,395 | | |
| | Even Yehuda | 458 | | |
| | Qedma | 964 | | |
| | Tel Mond | 218 | 9,040 | 1,300 |
| Herzliyya | Herzliyya | 4,817 | 4,820 | 700 |
| Kefar Sava | Ra'ananna | 1,411 | | |
| | Kefar Sava, | | | |
| | Hod HaSharon | 5,402 | | |
| | Ramat HaSharon | 2,368 | 9,250 | 1,315 |
| Dan | Dan Association | 20,000 | 20,000 | |
| | of Towns | 50,000 | 50,000 | — |
| Ayyalon | Or Yehuda | 891 | | |
| Association | Kiryat Ono | 1,663 | | |
| | Yehud | 977 | | |
| | Be'er Ya'aqov | 462 | | |
| | Ramla | 3,166 | | |
| | Lod International | | | |
| | Airport, Aircraft | | | |
| | Industries, Beit | | | |
| | Dagan, etc. | 2,877 | | |
| | Dagan, Tser'fin, | | | |
| | Kefar Habad | nil | 11,940 | 1,700 |
| Ashod | Ashod | 7,381 | | |
| | Yavne | 3,205 | | |
| | Gedera | 470 | 10,500 | 1,500 |
| Ashqelon | Ashqelon | 5,538 | 5,540 | 800 |
| Jerusalem | Jerusalem | 27,011 | 25,000 | 3,000 |
| Qiryat Gat | Qiryat Gat | 3,206 | 2,600 | 350 |

Source: Unpublished prefeasibility reports by Tahal.

TABLE 6.11

*Selected Data for Planning of Five Regional Schemes (1985)*

| Scheme | Annual Flow MCM | Cost of Disposal with No Utilization Scheme Ag/cu.m | Summer Utilization MCM | Cost of Summer Utilization Ag/cu.m | Winter Disposal MCM | Cost of Winter Disposal Ag/cu.m | Notes |
|---|---|---|---|---|---|---|---|
| Natanya | 9.0 | 9.6 | 3.8 | 11.6 | 5.2 | 12.1 | Including treatment ponds of Beit Yitzhak, Beit Herut and Even Yehuda |
| Southern Sharon | 14.0 | 9.1 | 5.9 | 16.5 | 8.1 | 11.2 | |
| Ayyalon Assoc. | 11.9 | 13.8 | 5.2 | 17.4 | 6.7 | 13.7 | Disposal near Dan Project according to Tahal plan |
| Jerusalem | 26.0 | 8.7 | 11.0 | 13.4 | 15.0 | 13.6 | Disposal into the sea through special line from Yesodot or Qedma |
| Dan-Ashdod | | 6.5 | 11.5 | 26.0 | 2.0 | 8.2 | 11.5 MCM made up of 3.5 MCM of Ashdod and Yavne effluents and 8.0 MCM of Dan Project effluents. |

Source: Unpublished prefeasibility reports by Tahal.

## The Institutional Organization of Israel's Water Economy and its Relation to Re-Used Wastewater as a Supply Alternative

The institutional character of water resources in Israel includes administration, law, and public policy which is substantially different from that of any state or region in the United States. The extent of this difference is, in fact, almost polar. In the United States very often the preemption of increments in technical efficiency to the supply system is often the result of past piecemeal solutions to immediate problems.[14] As opposed to the fragmentation of authority and jurisdictions that occurs in the United States, Israeli policy, thanks to the conditions set by the Water Law of 1959, is highly cohesive. The stepwise integration of the water system in attenuated stages of development has avoided many of the problems of option preemption. The necessity of instituting laws for interregional water transferal arose from a

14. See, for example, National Academy of Sciences, *Water and Choice in the Colorado Basin*, NAS Publications, 1689, Washington, D.C., National Academy of Sciences, 1968); J. Bain, R. Caves, and J. Margolis, *Northern California's Water Industry* (Baltimore, Johns Hopkins Press for Resources for the Future, 1966); L. Zobler, et al., *Benefits from Integrated Water Management in Urban Areas—The Case of*

*the New York Metropolitan Region* (Alexandria, Va., National Technical Information Service, 1969).

15. A more complete description of administrative aspects of water use in Israel can be found in O. Tamir, "Administrative and Legal Aspects of Water Use in Israel," in *Water in Israel*, Israel, Ministry of Agriculture (Tel Aviv, 1973), Part A.

need to amend the constraining Ottoman Code of Law which forbade even intercommunity transfers. Partial solutions that circumscribed Ottoman Law were inadequate to the late 1950's. Only the comprehensiveness of the Water Law of 1959, making water resources of the State public property under the responsibility of the Minister of Agriculture, could provide a substantial enough departure from Ottoman Code. In practice, the Water Commission administers the law.[15]

The Water Commissioner, subject to the parliamentary responsibility vested in the Minister of Agriculture, controls all aspects of administration. The Water Law has established a system of administrative and quasi-statutory committees to cover all aspects of the supply and planning of water resources: the Hydrological Service which supplies hydrological data to the Commissioner; the Department for the Efficient Use of Water, which takes care of water metering, prevention of water pollution, domestic wastes, etc; and the Water Allocations and Licensing Department, which administers the laws for water-withdrawal rates, recharge, and the supply of water in general. This department works closely with agencies in the planning for the supply of water, such as Tahal Consulting Engineers, and Mekorot, the national water company that supplies some 60 percent of total withdrawals. The Department has a Surface Water Committee, Boreholes Committee, Committee for the Use of Marginal Waters, Water Allocations Committee, Recharge Committee, and Water Supply to Agriculture Committee. The Drainage and Soil Conservation Department functions to prevent floods and is responsible for drainage of agricultural land and storm runoff; and the Legal Bureau provides counsel to the Commissioner on relevant matters. The Economic Bureau collects data for water-planning purposes.

The Water Law has established, primarily for political purposes, a number of administrative and quasi-statutory bodies through which the public may enjoy wide and comprehensive participation in decision-making. This group of bodies should permit public participation in the National Plan for wastewater reclamation. The coordination of the various departments cited above, along with the sensitization of public officials through the vehicle public bodies, is assuring the successful implementation of the Dan scheme as well as the National Plan. Naturally, a good deal of intracommunity resistance to projects (because of odors from aeration ponds, etc.) cannot be resolved by national structures. Indeed, the Dan scheme did create opposition by residents of Rishon Le-Zion to certain locations of ponds. On the other hand, given the constraining conditions of supply in Israel, the public appears not to be highly resistant to the technology itself.

*Public acceptance in Israel.* Unlike the United States, the question of public acceptance in Israel is one that requires hindsight rather than foresight, since the implementation of the National Plan is already under way and the Dan scheme has been proceeding for several years. The public has paid little attention during the Dan scheme issue, and policy has been carried out as a matter of course with little resistance from officials themselves. This contrasts with the general tendency of United States public officials and managers to overestimate the unwillingness of the public to accept re-used wastewater. (The reasons are discussed at length in Chapters 8 and 9.) The public has welcomed the innovation that water courses (wadis) serving as open sewers will undergo treatment, thus eliminating visual and olfactory pollution and nuisance from towns and conservation and recreation areas.[16]

Since the costs of re-used wastewater for potable supply are somewhat higher than those of existing supplies, the question remains whether the populace is willing to pay for this new technology as a potable supply. Darr, Feldman, and Kamen indicate a willingness to pay more for supply.[17] Their study administered to some 1,892 residents of the main urban areas of Israel a battery of items on their willingness to pay more for water of better quality, for continuous supply (noninterruptible service), and for unrestrained unlimited use. In aggregate the answers to the questions indicated that residents were willing to pay more for supply at a level near that of the additional cost for re-used wastewater as a potable supply. If one assumes that the question of the public's willingness to consume reclaimed wastewater is resolved, then it appears that the demands for the product are sufficient to merit increments in the capacity of this technology.

The chapter to follow will indicate some correlation of public attitudes toward re-using wastewater with socioeconomic indicators, previous knowledge of and experience with reclaimed wastewater, the future inadequacy of conventional supplies, and the adequacy of the quality of the present water source. The present author examined these predictors of acceptability to re-used wastewater under extreme environmental conditions.

Geographers have already pointed out that man's adjustment to extreme environments or environmental conditions may not be defined by the usual sociodemographic or economic criteria.[18] Thus residents of an extremely arid environment may not behave the same as those in more humid areas. The extreme case of Eilat, a Red Sea port and resort in the extreme southern portion of the Negev Desert in Israel, merits an in-depth study.

Originally established in the early 1950's for its port function, as well as to secure Israel's shipping access to Indian Ocean trade, Eilat by 1972, only twenty years after its founding, boasted a population of over 14,000. The town's stable climate makes it attractive as a winter resort for the Israeli populace and as a summer vacation place for western Europeans; daily mean cloudiness averages less than 20 percent. The town has an average of 25 mms. of rainfall annually and summer mean maximum monthly temperatures approaching 40° centigrade. A highly mineralized aquifer some 40 kilometers north of the city supplies Eilat with fresh water for about 50 percent of its withdrawals. Two desalinization plants, with a total capacity of 6,000 cubic meters daily, supply the other 50 percent. Despite blending of the waters, overall water quality still remains poor.[19]

The author conducted a sample survey in Eilat of seventy-four residents stratified according to neighborhood. The following questions (translated from the Hebrew) were used to solicit public attitudes toward acceptance of reclaimed water:

(1) Would you drink wastewater that has been treated to the point of potable supply?

16. This is a general conclusion reached by the author during a two-year period studying alternatives in water management within Israel.

17. Darr, Feldman, and Kamen.

18. See the excellent review and bibliography in Gilbert F. White, *Natural Hazards* (New York, Oxford University Press, 1974).

19. Eilat residents expressed more concern about water supply quality than did the residents of nine other urban places in Israel.

20. Kally, "Israel's
Water Economy."

(2) Would you drink wastewater that has been treated to the point of
potable supply if the Ministry of Health approved of it?

*Yes* or *No* responses:

| Question 1 | | Question 2 | |
|---|---|---|---|
| Yes: | 22 | Yes: | 54 |
| No: | 28 | No: | 8 |
| Don't Know: | 24 | Don't Know: | 12 |
| Total | 74 subjects | Total | 74 subjects |

Contrary to the findings of the United States studies reviewed in the next
chapter, the responses to the two questions were not correlated with the
respondent's education, income, or previous knowledge of the technology.
The answers to Question 1, however, are similar percentage-wise to the
findings of most attitude surveys in the United States. Whereas one might
have expected greater receptivity in Eilat than in the American context, it was
not evident.

It appears that the populace puts a great deal of faith in the prospect of
re-used wastewater as a potable supply as long as the government puts its
official stamp of approval on the process (Question 2). Yet, the extreme
environment may be responsible for the lack of association between what
are normally valid predictor variables and attitudes toward wastewater re-
use.

The additional increment to supply in Eilat is in the form of a desalinization
plant now under construction. The marginal cost of supply from this plant is
approximately U.S. $2.50 per thousand gallons, far above any estimate for
potable reclaimed wastewater. The present author believes that the adjustment made by public officials in Eilat was merely a product of the existing
momentum of desalinization as a technology already in use in Eilat rather
than a response to purely economic considerations. Additions to and modification of existing wastewater treatment processes would surely have been a
more effective solution.

## Economic Aspects of Re-Used Wastewater in Israel

According to Kally, urban effluents in Israel have a 50-percent utilization
factor resulting from reclamation.[20] This relatively low recovery figure is attributable to four factors:

(1) Part of the water is lost through evaporation.
(2) Industrial effluents are highly contaminating—to the extent that their
purification becomes impracticable and their disposal must occur
on-site or before they can gain access to the domestic sewage.
(3) Israel's wet-dry, winter-summer climate prevents maximum use of
winter effluents which would require either large storage facilities or
ultimate disposal in an unused state.

(4) Effluents are available in highly populated regions distant from areas requiring irrigation, but conveyance and transmission facilities are extremely costly and not always economically feasible.

Albeit the foregoing conditions are highly restrictive, Kally does admit that the effluents will be adequate in volumes to compensate agricultural deficits arising from expanding domestic and industrial sectors.

It should be pointed out that whereas the above problems are relevant to the country as a whole, they, with the exception of the first point, are not likely to be severe in the Negev, particularly around Beersheva. In this region the effluents can be utilized year round, since little winter rainfall occurs anyway. Also, the effluents would not jeopardize the ground-water regime, as in the densely populated northern or central areas of Israel where wastes not treated fully (as in the Dan River Scheme) contain nitrates that may contaminate ground water used for drinking purposes.[21] The primary disadvantage of the Negev scheme, of course, would be conveyance distance from central Israel.

*Some "pure" economic considerations for irrigation water.* The competition of different qualities of irrigation water is such that decreasing crop yields resulting from reclaimed wastewater must be weighted against higher-priced fresh water. Decreasing crop yields would occur from the high chlorides content of the "secondary" treated wastewater. Whereas the elimination of these chlorides through further treatment would diminish the noxious effects of wastewater, the cost to do so would be considerably higher. A model[22] of the trade-off between the waters of different quality and treatment level (or no treatment) can illustrate the profitability of those levels for given crops. Assume

(1) the existence of homogeneous physical characteristics in a region (isotropic surface);
(2) given crop prices;
(3) an absence of pecuniary or technological external economies;
(4) perfectly elastic supply of waters of different quality;
(5) an absence of legal, sanitary, or institutional constraints.

Given the foregoing assumptions, if the objective function is to maximize profits, the next step is to determine the optimal level of treatment. It is assumed that sustained production occurs over the long run.

The farmer can calculate his average productivity function and his marginal productivity function to show the response of crop growth and yield to successive increments of water of certain qualities. It is possible to compute the net value of average product and net value of marginal product of that water quality. By deducting, from the successive values of yield that are attributable to a particular water quality with non-land factors, the costs of the other non-land factors used with the specified quality water increments, these net values may be calculated for different water qualities. The resultant functions would be net value of average product (VAP) and its corresponding net value of marginal product (VMP).

21. The concentration of nitrates in drinking water is the etiology for the disease of methemoglobinemia in infants.

22. Adapted from Bain, Caves, and Margolis, Appendix A.

23. Indeed a number of researchers within Israel at Tahal and the Hebrew University School of Agriculture have calculated production functions for saline water. For example, Darr has used the "Average Residual Value" method where:

$$MVP_m = \frac{C - \sum\limits_{i=1}^{m-1} W_c A_i}{A_m}$$

C = value of agricultural product
$W_c$ = MVP of Factor
       (i)i=1, 2, or m−1)
$A_i$ = input level of factor
$MVP_m$ = MVP of saline water

See P. Darr [Dalinsky], "Intersectoral Competition for Brackish Waters" (unpublished Tahal memorandum, Tel Aviv, 1972).

The farmer can then calculate these functions for waters of different quality and for each crop under consideration. Comparisons would yield:

(1) the quantity of water of each quality per unit of land for each crop needed for maximum VAP;
(2) the height of the VAP function itself for each water quality and each crop;
(3) the shapes of VAP and VMP curves for each water quality and crop over a larger range of water quantity applied.

The general axiom for decision-making would be that, to maximize VAP for a particular crop, the quality of water which uses the same quantity of water per acre as another quality and has a lower maximum VAP will tend to be absolutely inferior. In turn, a calculation for specific crops should be made. The criterion for choosing each crop should be that a crop which maximizes VAP in the same quantitative application of a specific water quality as another crop and has a lower maximum VAP will tend to be less desirable.

Crops can then be chosen on the basis of successively decreasing water requirements (once the specific water quality has been determined) and successively increasing maximum net values of average product.[23]

*Comparisons of the costs of water production.* Israel's cost-accounting structure of present facilities poses a problem of comparing costs of water supply for different production alternatives. Given the real cost structure of the facilities, the use of estimates only allows one to approach ranges in costs for particular technology. Table 6.12 shows such approximations, which have been compiled by the author from a number of sources. Clearly, wastewater reclamation and artificial rainfall stand out as the least costly alternatives.

Prices in Israel do not reflect costs because the Water Commission's Equalization Fund subsidizes the high cost of supplying consumers located in central and southern Israel. Recent legislation has related average costs to prices, but this does not obviate welfare losses incurred by the departure from marginal cost pricing. Indeed the rapid growth in water use in the total economy reflects not only increases in income and population but also the nonoptimal pricing policies.

*Other economic considerations.* With the consequent projected addition to the economy of some 270 million cubic meters of water from reclamation by 1985, one can infer certain pecuniary externalities. That is, would such additions create price distortions because of extra crop output through additional irrigation water or an impact on the prices of the factors of production? First, the increments in supply would merely account for what would have been the deficits in total water production by 1985. That is, the reclaimed water for irrigation would replace agricultural cutbacks (transfers from agricultural to urban use) or aquifer "mining." This quantity would be almost equal to the growth in water demand of the industrial and urban sectors by 1985. During this time period agricultural production would probably increase per unit of

TABLE 6.12

*Estimated Average Water Costs for Various Production Alternatives*
(in U.S. dollars per 1000 gallons)

| Alternative | Cost in $U.S./ 1000 gallons |
|---|---|
| Desalination (large-scale) of seawater | 2.20–3.00 |
| Wastewater reclamation (secondary treatment only) | .20–1.00 |
| Artificial rainfall (including necessary infrastructure) | .05– .20 |
| Conventional supply–national water carrier | .08– .16 |
| Desalination of brackish water (small-scale) | 1.50–3.00 |
| Wastewater reclamation (Dan scheme) | 26.– 40 |
| Stormwater drainage and deep-bore wells | .50–2.00 |

24. D. Yarden, *Drought Compensation Payments in Israel*, Natural Hazards Research Working Papers, 24 (Boulder, Institute of Behavioral Studies, University of Colorado, 1974).

water applied, because of improvements in technology, presumably at least enough to keep up with domestic demands for crops and to maintain Israel's role in supplying external markets. Given a *ceteris paribus* assumption about crop mixes and water prices, under the above conditions pecuniary externalities will be minimal. On the other hand, if additional irrigation water were made available by assuaging the effects of drought, compensation payments in Israel could potentially be reduced. These payments are based upon the costs of crop production.[24] For the period from 1962 to 1969 some I£36,000,000 have been paid through the Ministry of Agriculture to farmers in the designated drought compensation zones. In the thirteen years previous to that period, the government expended some I£45,000,000. The addition of supplementary irrigation facilities to the areas eligible for compensation could very likely defray the bulk of these monetary transfers.

## Conclusions

Israel's water-planning agencies have successfully dealt with the specter of natural hazard despite physical conditions that are adverse to rapid economic development. But the immediate outlook appears to be dimmer. Increasingly, irreversible processes stemming from a lack of water-pollution control and supply facilities are encroaching, narrowing the range in choice of alternatives for planning. Reclamation of wastewaters, through secondary and tertiary processes, can provide quantities of supply at necessary quality to make up the shortfall needed for expanding urban development while maintaining Israel's agricultural self-sufficiency and export potential. Already, these options are in operation on a large scale and proposed expansion is under way. Indeed, the case of Israel may provide an analog for other regions that are experiencing or will experience similar water supply and pollution problems.

**Stephen Feldman**

The Southwest United States is one of those areas in which these problems will be encroaching upon efficient resource utilization. The Israeli paradigm may be of practical use to planners in that area, since the problems of Israel in the 1950's are becoming evident in the Southwest of the 1970's.

# Public Acceptance of Water Re-Use: A State-of-the-Art Analysis

Roger E. Kasperson

# 7

Unlike Israel, the United States has enjoyed a plentiful water supply at bargain prices. Despite some difficulties noted in Chapters 1 and 4, this water has also been of a generally high quality, and thus free of the water-borne diseases of the nineteenth century. It is not surprising, then, that Americans are satisfied with the water they get from the tap. A 1973 Gallup Poll commissioned by the American Water Works Association revealed that 83 percent of Americans feel they have enough water for their needs and 70 percent are generally satisfied with its quality.[1] There is also a high regard for the guardians of the supply—the water managers and public health officials who have built the water-supply systems, set the health standards, and monitored the product delivered to the public. Until recently, there was little to disturb this happy state of affairs.

The "great fluoridation controversy" had exposed many water managers to rancorous community conflict. Accustomed to an orderly administrative process of water-supply augmentation, these managers suddenly found themselves squarely in the middle of heated battles which frequently turned on issues unrelated to water-supply or public-health considerations. Now, before these clouds have even cleared from the horizon, many water managers see a new storm brewing. The prospect of incorporating treated wastewater directly into the potable water supply shocks many managers and augurs a new round of public controversy.

Henry Graeser, Director of the Dallas Water Utilities Department, believes that the reluctance of water managers to face squarely the water re-use question can be attributed to the availability of other, less costly, high-quality water sources; the research lag in quality control for re-using water; and "the fear of violent public reaction at the thought of any used water entering the primary source."[2] In his study of ten Philadelphia Water Department officials, James Johnson found that all believed that the community reaction to a proposal for renovated wastewater would be disapproval. He concluded: "it would appear that water managers know very little of consumer responses concerning renovated wastewater, but generally consider that the public would not accept it . . ."[3] An extensive study of 158 federal, state, and local water resource officials in 1973 found that these public officials "grossly underestimated public acceptance."[4] Whereas 25 percent of the public officials were willing to accept renovated wastewater for drinking, they estimated that only 1.4 percent of the public would be similarly disposed.[5] These few examples could be replicated many times over. There is ample evidence that

1. Gallup Poll, "Water Quality and Public Opinion," *Journal of the American Water Works Association*, 65 (1973), 513–519.

2. Henry J. Graeser, "Water Reuse: Resource of the Future," *Journal of the American Water Works Association*, 66 (1974), Pt. I, 576.

3. James F. Johnson, *Renovated Waste Water: An Alternative Source of Municipal Water Supply in the United States*, University of Chicago, Department of Geography Research Papers, 135 (Chicago, 1971).

4. Ralph Stone and Company, Inc., *Wastewater Reclamation: Socio-Economics, Technology, and Public Acceptance* (Springfield, Virginia, National Technical Information Service, 1974), p. IV-4.

5. Ibid., pp. IV-5, IV-6.

water managers see public acceptance as a major obstacle to high-order uses of reclaimed wastewater.

The chapters to follow will discuss in detail the reasons for this view of public acceptance. Suffice it to note here that with professionals themselves deeply concerned over safety issues and reluctant to embrace what amounts to a revolution in familiar means of supply augmentation, deep opposition by the public would be comforting. Perhaps the fluoridation ghost still stalks the corridors of water managers, or perhaps officials simply are projecting their own reservations onto the public. On the other hand, perhaps it is the professional's greater knowledge and understanding. Whatever the roots of this perception, there is a need to test it against reality. Given the dearth of actual experience with water re-use accumulated to date and a growing number of public acceptance studies, what is the present state of knowledge on this subject? Delineating this problem—an undertaking more formidable than it might initially appear—is the task of the present chapter.

## An Elusive Problem

For all the visibility of the environmental movement over recent years, the notion of water reclamation and re-use is still new to most people. The formation of attitudes requires time—to search for and sift information, to talk over issues with family and friends, to gauge the opinion of experts. For most Americans the water re-use prospect has not been of sufficient immediacy to allow for this "mulling over" process. Attitudes to reclaimed water are only partially formed and the issues involved only partially understood. Thus gauging the public response to water re-use projects requires both cleverness in research design and caution in interpreting results.

*Asking the right questions.* The first hurdle entails the manner of soliciting attitudes. Because information is often rudimentary and attitudes still in an embryonic stage, the researcher faces a difficult choice. If he solicits directly the respondent's reaction to using reclaimed wastewater, the interviewer runs the risk of eliciting a viewpoint which may not correspond at all to what that attitude will be when the same respondent, in the future, actually confronts a choice. To solicit an attitude yet unformed is, of course, a procedure logically flawed from the onset. Hence a number of researchers have felt obligated to provide a definitional or brief informational statement of water reclamation and re-use.

But there is unlimited opportunity for mischief in such a procedure. Since many of the investigators of public acceptance are proponents of re-use, the phrasing of information has sometimes reflected their bias. Moreover, even where researchers have tried to mirror accurately a description of water re-use free of emotive signals, they still may fail to simulate the informational and emotional environment in which the public may find itself. A scientifically accurate definition of re-use does not capture the division of opinion which may be expected if controversy erupts. So much depends upon the environment one expects to prevail when public opinion forms and matures.

Much also rides upon information provided and the phrasing of questions aimed at soliciting attitudes, particularly the key attitude (dependent variable) that one seeks to explain. Consider the likely response to be expected from:

"Would you object to drinking recycled sewage?"          (Gallup Poll)

compared with

"Would you drink water of a quality equal to your present supply if you knew that it was renovated waste water?"          (Johnson Study)

Given the difference in the language employed in the interview item, it is not surprising that the first resulted in public acceptance at a 38.2 percent rate whereas the latter found 77.4 percent favorably inclined.

To test whether the phrasing of questions did indeed produce different levels of public acceptance, the surveys conducted by Clark University during 1970 and 1971 purposefully employed two different informational statements. The first, similar to that used in other studies, presented a precise and sanguine definition, free of risk:

Reclaimed water is municipal sewage which has been cleaned and treated. Such water would be used to add to the present water supply. Reclaimed water meets the standards for drinking water of the U.S. Public Health Service and can be produced at a reasonable cost. It has been used for fishing and swimming in Santee, California and for drinking in the industrial town of Windhoek, South West Africa. Because it improves the quality of waste water returned to rivers and lakes, water reclamation also helps to reduce water pollution.

The second version added a sentence that gently introduced the element of risk:

Some authorities, however, feel that science is not yet advanced enough to determine whether there is a health hazard involved.

Public response to the two versions was significantly lower (significant at .01; that is, there is one chance in a hundred that the difference is due to random variation) for the second when use of the reclaimed water involved ingestion. This little experiment calls for a searching look at the survey questions and language before the results can be taken at face value. Many of the seeming differences in findings are, in short, functions of the way in which questions are asked.

*Asking the right people.* No less difficult is the decision on whom to interview. The limited state of public knowledge encourages attention to communities where water re-use is a legitimate part of the planning environment. And the bulk of the public acceptance surveys has in fact been conducted in arid and semi-arid areas of the West and Southwest. Indeed, there has been a tendency to focus on communities with successful water re-use experiences. This orientation to survey research is undoubtedly both necessary and appropriate but also suggests that emerging findings may present an overly optimistic view of the threat of public controversy and rejection. The

6. Robert L. Crain, Elihu Katz, and Donald B. Rosenthal, *The Politics of Community Conflict* (Indianapolis, Bobbs-Merrill, 1969).

7. William Gamson, "The Fluoridation Dialogue: Is It Ideological Politics?," *Public Opinion Quarterly*, 26 (1962), 526–537.

study of public acceptance in Israel discussed in the previous chapter provides collateral evidence on this issue. Ideally, research over time would attempt to balance interviewing among (1) communities that had accepted and rejected water re-use projects, (2) areas of the Northeast where protected, single-purpose upland reservoirs are the preferred norm, and (3) communities of the water-short West and Southwest.

*Substituting for experience.* Perhaps the most significant problems pervading the study of public acceptance, however, are the newness of planned water reclamation and re-use technology, particularly for high-order uses (for example, swimming, drinking), and the lack of community adoptions of such innovations. Put simply, there is a need to predict public response when there are no examples to analyze. As a result, researchers must resort to surveys, projections, and analogies in lieu of a record of experience. This is a tricky business. Authoritative statements on public acceptance must, as a result, await the first set of pioneering communities where water re-use will arise from local initiative and not from federal demonstration projects.

With these caveats firmly in mind, it is appropriate to turn to the major analogy available for water re-use decisions and to consult the body of studies and experience which comprise the present state of knowledge on public acceptance.

## Fluoridation: How Valid is the Analogy?

The fluoridation of public water supplies began in 1944 in Grand Rapids, Michigan, where fluorides were added in an effort to combat tooth decay. By 1951 the U.S. Public Health Service and the American Medical Association, convinced by the dramatic achievements of fluoridation in the early adopting cities, formally endorsed the innovation. The rest of the scientific and professional community soon followed suit. But by 1952 a national network of opposition had formed, and the rate of community adoptions was already in decline. In community after community the local public health officials, doctors, dentists, and political elite all endorsed the proposal, only to find it defeated at the polls. Even more puzzling, the successful opponents were professionally marginal people—little-known in their communities, older, lacking status and education, and not previously active in other local political issues.[6] The chief reason, as William Gamson has made clear, is fear of medical side effects and not, as some have suggested, socialism or big government.[7] Fifteen years after the first successful adoption and after the assemblage of nearly universal support in the scientific and professional community voters in two out of three public referenda were systematically rejecting fluoridation. Most recently, Los Angeles voters in 1975 rejected a proposal to fluoridate the city water supplies.

The fluoridation experience also attests to the importance of executive leadership and successful innovation by the community. One of the central findings by Donald Rosenthal and Robert Crain in a comprehensive study of

496 fluoridation decisions in 362 cities was that the willingness of the mayor to support openly the innovation virtually determined the outcome of the community decision.[8] If the mayor did not support fluoridation or adopted a neutral stance, fluoridation enjoyed very little chance of success. In addition, if the executive leadership dodged the issue, and sent it to referendum, a favorable decision was remote.

There may be several important lessons here for water reclamation and re-use. First, whereas many observers have characterized fluoridation as an example *par excellence* of an irrational public in operation, such a view, this author would contend, makes sense only through the manager's or scientist's prisms. Given his state of knowledge, a favorable decision was indeed rational. But the public saw an innovation that promised only a marginal reward—reduced tooth decay for children—obtainable in other ways but which conceivably could involve serious medical hazards. Since the fluoridation opponents could seemingly cite medical and public health experts who pointed to important risks, the public, beset by contradictory information and expertise, cautiously decided against innovation. Fluoridation suggests that a divided medical or public health community could incite particularly strong adverse public reaction in the case of direct potable use of renovated wastewater.

On the other hand, there is a good chance that the fluoridation controversy may not be replicated even in the case of direct potable re-use. Fluoridation involved a marginal gain obtainable through voluntary means (for example, by fluoride pills); direct potable re-use will only be employed in the absence of other options, and it will involve a critical benefit—an adequate and safe water supply. Moreover, there is no evidence that water re-use decisions will not occur as routine administrative decisions augmenting water supply, although lack of public involvement and education could, as this chapter will make clear, carry significant risks.

But if the community's executive leadership does not see water re-use as a routine decision, chooses to spread the risk, and encounters division in the scientific and regulatory communities, then fireworks could indeed explode in the area of public acceptance.

## Survey Research: An Overview

Although survey research on attitudes toward water re-use is still in its infancy, a very useful record of studies is now emerging. The need exists, of course, to discriminate, for the quality of studies to date is very uneven. A brief overview of major findings may provide a useful backdrop from which to assess the present state of the art on public acceptance issues. An early 1965 telephone survey[9] in thirty-six communities in Texas, Kansas, Illinois, and Massachusetts, which defined re-used water to the respondents as water that had been used for either domestic and/or major industrial purposes and then filtered and purified, found a majority (62 percent) of the 722 respondents willing to drink re-used water (Table 7.1).

8. Donald B. Rosenthal and Robert L. Crain, "Structure and Values in Local Political Systems: The Case of Fluoridation Decisions," in *City Politics and Public Policy* ed. James Q. Wilson, (New York, Wiley, 1968), pp. 228–229.

9. Conducted by Duane Baumann, this survey is discussed in Duane D. Baumann and Roger E. Kasperson, "Public Acceptance of Renovated Waste Water: Myth and Reality," *Water Resources Research*, 10 (1974), 667–674.

10. William H. Bruvold
and Paul C. Ward,
"Public Attitudes To-
ward Uses of Re-
claimed Wastewater,"
*Water and Sewage
Works*, 117 (1970),
120–122. The same
study is also reported in
William H. Bruvold, "Af-
fective Response to-
ward Uses of Re-
claimed Water," *Ex-
perimental Publication
System*, Issue 3 (De-
cember 1969), Manu-
script 1170, pp. 1–12.

11. The author is in-
debted to John Sims of
George Williams Col-
lege for first calling this
to his attention.

12. See J. C. Baumer
and P. G. Robertson,
eds., *Reclamation of
Wastewater for Munici-
pal Water Supply* (Bal-
timore, Department of
Geography and En-
vironmental Engineer-
ing, Johns Hopkins Uni-
versity, 1973), chap. 4;
and Robert B.
Athanasiou and Steve
H. Hanke, "Social
Psychological Factors
Related to the Adoption
of Reused Water as a
Potable Water Supply,"
*Urban Demands for
Natural Resources*, ed.
Western Resources
Conference (Denver,
University of Denver,
1970), pp. 113–125.

TABLE 7.1

*Variations in Public Willingness to Drink Reused Water, in Selected States*

| | Yes | | No | | Don't Know | | |
| | No. | % | No. | % | No. | % | Total |
|---|---|---|---|---|---|---|---|
| Texas | 155 | 59 | 90 | 35 | 19 | 12 | 264 |
| Kansas | 72 | 72 | 18 | 18 | 10 | 10 | 100 |
| Illinois | 134 | 74 | 30 | 17 | 17 | 9 | 181 |
| Massachusetts | 82 | 46 | 79 | 45 | 16 | 9 | 177 |
| Total | 443 | 62 | 217 | 30 | 62 | 8 | 722 |

Source: Duane D. Baumann and Roger E. Kasperson, "Public Acceptance of Renovated Waste Water: Myth and Reality," *Water Resources Research*, 10 (1974), 671.

The affirmative responses were significantly different in Massachusetts, where the respondents were nearly evenly divided in their acceptance of drinking renovated wastewater. Unlike respondents in other states, where 60–75 percent of those interviewed would drink it, only 46 percent of the Massachusetts respondents were willing to ingest such water. Thus there is early preliminary indication of regional variation in consumer acceptance.

In a 1969 study William Bruvold and Paul Ward analyzed public acceptance of potential uses of reclaimed water among twenty-five respondents in each of two California communities, both of which had water reclamation projects.[10] They concluded that the amount of expressed opposition to particular uses of reclaimed water is directly related to degree of bodily contact and that expressed opposition drops off sharply below the use of reclaimed water for swimming. Plotting on a graph, however, places the critical break at the ingestion of the water and thus supports even more strongly their contention that it should be possible to initiate uses of reclaimed water up to the body contact involved in swimming before encountering widespread public opposition (Figure 7.1).[11] Based on these results, Bruvold and Ward recommend a step-by-step process of innovation with water re-use, like that employed at Santee, to gain maximum public acceptance.

The Clark University pilot study of 220 respondents in the cities of Gloucester (Massachusetts), Wilmington (Delaware), Indianapolis, and Kokomo (Indiana) during the summer of 1970 provided some weak evidence of regional differences in public acceptance of reclaimed water. More significant was the remarkably consistent pattern of response to varying uses of reclaimed water (Figure 7.2) among the four communities, confirming the earlier Bruvold and Ward results. The research also suggested the relatively low prevailing knowledge of re-use, with between 29 and 31 percent of respondents in each site describing themselves as familiar with the idea. This may explain in part the relatively low acceptance (no more than 32 percent in any site) of the notion of direct potable re-use.

In Baltimore County, Maryland, a team from Johns Hopkins University conducted a simple random telephone survey of some 291 households in which attitudes toward direct potable and recreational uses were tested.[12]

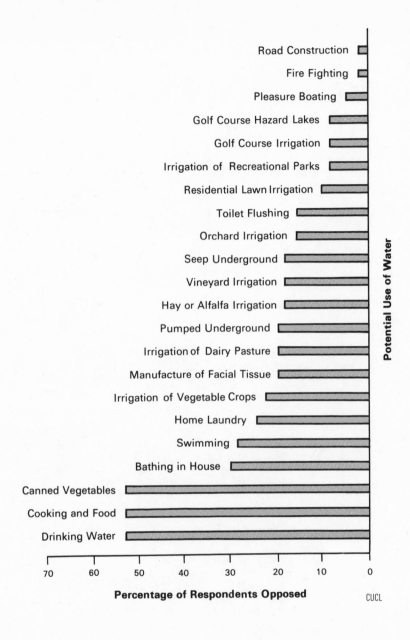

Figure 7.1: Attitudes toward Uses of Reclaimed Water.

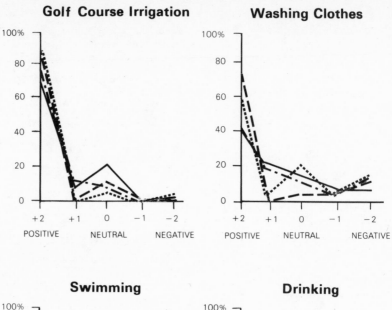

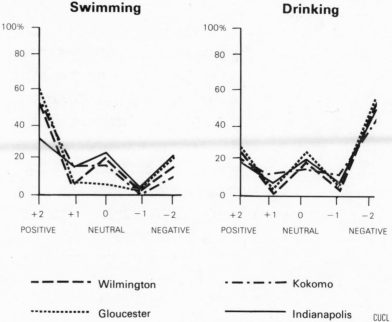

Figure 7.2: Attitudes toward Re-Use of Wastewater (by site).

Only education emerged as a strong indicator of acceptance for drinking, while sex and occupation were more weakly correlated. The authors predicted that early innovations with high-order uses of reclaimed water will have greatest success in high-status communities populated by young educated persons with relatively high incomes.

A survey conducted at Park Forest, Illinois, in 1971 found a startling 81 percent of respondents willing "to use recycled water in and around (their) home, if it was as good in quality and as safe bacteriologically as the water (they) are now using."[13] Here the high acceptance rate undoubtedly reflects the framing of the question, which, among other things, fails to specify whether ingestion is involved. If the respondents assumed nonpotable uses, then the results are consistent with other studies. The researchers also found that cost and type of re-use system (indirect versus direct) were significant factors in public acceptance.

In 1971, in one of the more significant studies of public acceptance, James Johnson surveyed some 221 persons in five cities: Philadelphia, Camden, Cincinnati, Tucson, and Portland, Oregon.[14] One of his most interesting findings revealed that 170 of his 221 interviewees would be willing to drink renovated wastewater, although again the phrasing of the item encouraged a high positive response. As did the Johns Hopkins University study, Johnson's research demonstrated that consumer acceptance increases as educational levels and knowledge of renovated water increase. In addition, cities with adequate future supplies or poorer quality water were more receptive to the adoption of reclaimed water for high-order uses.

The Clark University study proper, largely concentrated in areas of the Midwest, arid Southwest, and West revealed several important facets of attitudes toward water re-use. Involving some 408 lengthy personal interviews in the cities of Lubbock and San Angelo (Texas), Santee, (California), Colorado Springs (Colorado), and Kokomo (Indiana), the research highlighted the importance of crisis and prior experience with re-use.

Lubbock, Texas, which had an adequate existing municipal supply but was considering reclaimed wastewater for recreational use, provided the most decidedly negative response. Yet the cities that had experienced or were experiencing crisis situations in which wastewater re-use had been widely discussed as a supply alternative (San Angelo and Kokomo) or had accumulated experience with water re-use (Colorado Springs and Santee) clearly revealed the strongest favorable evaluation (Table 7.2).

It is particularly noteworthy to find that successful community experience with reclaimed water over time may well contribute significantly to enlarging public acceptance within the community. The better than two-to-one public acceptance of renovated wastewater for drinking among the sample in Santee provides a striking demonstration of the effect of the unique experiment in water re-use on public attitudes.

The Clark University study also suggests that, concerning the question of renovated wastewater, the reference group that the public valued most highly was the medical profession (Table 7.3). The respondents gave greater weight to the stand of United States public health officials, doctors, engineers, and scientists than to politicians and the media.

13. Albin D. Pagorski, "Is the Public Ready for Recycled Water?," *Water and Sewage Works*, 121 (1974), 108–109.

14. Johnson, chap. IV.

Roger E. Kasperson

15. See William H. Bruvold and Henry J. Ongreth, "Public Use and Evaluation of Reclaimed Water," *Journal of the American Water Works Association*, 66 (1974), 294–297; and Ralph Stone and Co., p. II-22.

16. Bruvold and Ongreth. See also William Bruvold and Paul C. Ward, "Using Reclaimed Water—Public Opinion," *Journal of the Water Pollution Control Federation*, 44 (1972), 1690–1696.

TABLE 7.2

*Community Variations in Public Evaluation of Renovated Wastewater for Drinking*
(percentages)

| Site | Evaluation Positive | Neutral | Negative |
|------|---------|---------|----------|
| Lubbock, Texas | 16 | 18 | 66 |
| Indianapolis, Indiana | 25 | 20 | 55 |
| Wilmington, Delaware | 27 | 18 | 54 |
| Gloucester, Massachusetts | 31 | 13 | 57 |
| Kokomo, Indiana | 38 | 2 | 61 |
| Colorado Springs, Colorado | 45 | 7 | 49 |
| San Angelo, Texas | 56 | 6 | 38 |
| Santee, California | 65 | 6 | 29 |

In addition, the data support the association that Bruvold and Ward found between public acceptance and the particular use of renovated wastewater. There is a clear step-by-step progression of resistance to the use of renovated water as one moves through the hierarchy of uses from nonbodily contact, to bodily contact, and finally to ingestion. Caution is still advisable, since the most recent research indicates considerable variance of acceptance within classes of use.[15] Nevertheless, the relationship is important in community efforts to prepare the public for wastewater re-use adoptions.

In a sequel to his earlier research, Bruvold in 1972 analyzed public acceptance by selecting pairs of communities similar in size, location, and socioeconomic characteristics but different only in that one (the "project" town) had an operative wastewater reclamation and re-use system and the other did not.[16] Despite a lack of significant differences in overall attitudes between the matched pairs of *communities* in acceptance of reclaimed water for drinking, there was a tendency for *individuals* who used reclaimed water to be more favorably inclined to drinking such water. The more meaningful discriminating correlates were education and belief about the relative merits of technical versus natural purification methods. Noting, however, that 50

TABLE 7.3

*Public Preference of Reference Groups*
(percentages)

| Reference Group | Most Preferred |
|-----------------|----------------|
| Doctors | 40 |
| Public health officials | 34 |
| Engineers and scientists | 21 |
| Mayor or city manager | 3 |
| City councilmen | 2 |
| TV and newspapers | 0 |

percent of the respondents did not recommend using reclaimed water for purposes involving personal contact, the study concluded—perhaps somewhat hastily, given its scope—that "the public is not yet ready for intimate uses of reclaimed water . . . attempts to employ these uses would very likely stimulate public protest seeking to terminate such innovation."[17]

For a 1972 M.S. thesis at the University of Colorado on public acceptance Robert Carley interviewed some 447 persons in Denver. Given various supply options, Carley found only a 37.8 percent favorable response to drinking reclaimed water.[18] When asked, however, "Would you drink renovated wastewater if its quality was the same as your present house water?," the positive response rose to 84.1 percent, indicating the variability of acceptance depending upon presentation of the issue. Carley also found that sex, age, and education were all important factors in attitudes toward re-use.

A 1973 Gallup Poll, commissioned by the American Water Works Association, found that in a national sample of 2,927 individuals some 54.5 percent objected "to drinking recycled sewage."[19] Given the phrasing of the question, it is perhaps noteworthy that some 38.2 percent responded affirmatively. The poll also revealed that women, older adults, and the less educated were more resistant to direct potable re-use than other groups.

Finally, in research funded by the U.S. Office of Water Resources Research, Ralph Stone and Company conducted a telephone survey of some 1,000 residents of ten California communities.[20] Unfortunately, an unacceptably high rate of refusal to participate (31 percent) and the disadvantages of telephone surveys detract from the results. Research findings of a 39.1 percent acceptance of reclaimed water for various uses tended to follow the pattern described in the Bruvold and Clark University studies. In explanatory variables, only age emerged as a significant factor; all other correlates showed only weak associations.

These studies together comprise an important part of knowledge on attitudes toward water re-use. But it is equally, and perhaps more, important to assess what has occurred in the limited experience where water re-use has actually been tried.

## The Record of Experience

Chapter 2 discussed many examples of reclaiming water for such low-order uses as golf course irrigation, industrial cooling and boiler feed, and irrigation, while Chapter 6 examined briefly public acceptance in Israel. For high-contact uses of reclaimed water, such as swimming and drinking, there are very few examples. Three cases, however—Chanute, Windhoek, and Santee—provide useful information for analysis of experience to date.

In Chanute, because of the poor water quality involved, the public remained uninformed until after the recirculation had begun. Public reaction was negative, and consumers moved quickly to a variety of adjustments. Bottled water sales boomed as nearly all grocery stores stocked large supplies. Residents drilled more than seventy wells, most of which were too

17. Bruvold and Ongreth, p. 297.

18. Robert Lane Carley, "Wastewater Reuse and Public Opinion," unpublished M.S. thesis, Department of Civil and Environmental Engineering, University of Colorado (1972).

19. Gallup Poll.

20. Ralph Stone and Company.

21. Neil Ackerman, "Water Reuse in the United States," unpublished M.A. thesis, Department of Geography, Southern Illinois University (1971).

22. Johnson, p. 21.

23. E. W. Houser, "Santee Project Continues to Show the Way," *Water and Wastes Engineering*, 7 (1970), 40–44.

mineralized for domestic use. Also, many people hauled water from neighboring towns and wells. In a 1969 telephone survey of thirty-nine Chanute residents who experienced re-use in 1956–57, Neil Ackerman found, perhaps optimistically, that 61.7 percent thought that everyone eventually accepted drinking the treated wastewater. And 71.7 percent of the sample recalled the tap water as having been acceptable.[21] Although Ackerman's results may be unduly favorable (most managers at the time were convinced that the public rejected the water for drinking), the evidence suggests that even the poor quality product of the Chanute experience met at least some partial public acceptance, particularly for lower-order uses. Even those who refused to drink tap water most likely did use reclaimed water for flushing, washing clothes, and bathing. No discernible groups of opponents or proponents took up the cudgels in Chanute.

Public acceptance in Windhoek, by contrast, was apparently remarkably widespread, and water supply officials there reported nearly unanimous approval from various consumer groups. One of the interesting facets of the Windhoek experience involved the implementation of the re-use system. Not until three weeks after the changeover did the public learn of the addition of renovated water. Indeed, thanks to advance publicity, the prevailing opinion had it that city residents were already drinking reclaimed water. The new blend of renovated water and water from the existing supply came through noticeably improved in quality. It is scarcely surprising, then, that several inquiries asked whether Windhoek had abandoned the water reclamation system.[22]

At Santee the very design of the entire reclamation project courted maximum public acceptance:

> As to the gaining of public acceptance, the best approach appeared to be to overcome the negative connotations of sewage by associating the program with pleasant things that the public enjoys and approves. To accomplish this end, it was decided to put the reclaimed water in an attractive setting and invite the public to look at it, sniff it, picnic around it, fish in it, and swim in it. This approach would give great numbers of people an opportunity to examine the water and satisfy themselves as to its acceptability. At the same time, they would gain an understanding of the tremendous public benefits that could be created through the reclamation of wastewaster.[23]

Accordingly, managers adopted a careful sequence of public education and involvement. They made oral presentations to a variety of community organizations and they offered extensive explanation in the local press in advance of the project. They invited the public to view, through a chain link fence, the enticing waters of the completed lakes. When the landscaping of the recreational lake was completed in the spring of 1961, they allowed the public to use the park for picnicking, boating, and scenic beauty. Later, they inaugurated a fish-for-fun program, but fishermen could not take their catches home for eating before June 1964. Finally, in June 1966, they opened a pool using renovated wastewater for swimming. By the end of that year some 75,000 persons had visited the aquatic park, and by 1966 that figure had jumped to 125,000, many of whom expressed interest in the use of reclaimed water for drinking. Despite its recent demise, the Santee experience demonstrates convincingly that it is indeed possible to muster widespread public support for high-order uses of reclaimed wastewater.

24. Bruvold and Ward.

## Attitudes toward Reclaimed Water: A Comparison of Findings

Perhaps the single most telling attribute of public attitudes toward reclaimed water is the increasing resistance encountered with the more intimate uses. Bruvold and Ward first uncovered this ladder of attitudes.[24] In the 1971 five-city study conducted by Clark University, acceptance of reclaimed water was lowest (48–51 percent) for ingestion, jumped significantly (to 78–81 percent) for uses that involved body contact but not ingestion, and approached universal acceptance (94–96 percent) for noncontact uses. Table 7.4 summarizes these study results.

Four of the eleven studies of public acceptance have analyzed the difference in attitudes according to potential uses of the water. Figure 7.3 summarizes the findings of these four studies. When compared on the same graph, the findings reveal a remarkable consistency, conducive to an expectation of a "ladder of acceptance" related to the use of reclaimed water in most American communities.

As for the highest prospective use—drinking—there is less consistency in findings to date. The willingness to drink reclaimed wastewater uncovered in 10 of the 11 studies varies from a high of 81 percent to a low of 29 percent. A comparative summary of the findings appears in Table 7.5. Three studies—those of Baumann, Pagorski, and Johnson—reveal high levels of acceptance ranging from 62 to 81 percent. Interestingly enough, each of these three presented the issue to the respondent in a very optimistic way, and each was conducted in more humid regions where lower acceptance might normally have been expected. On the low side, the 29 percent acceptance of drinking uncovered in the Clark University pilot study is undoubtedly a product of the question, which presented choices on a three-point scale (favorable, neutral, unfavorable) instead of the normal binary scale. If the neutral responses are apportioned equally among positive and negative, the acceptance level rises to 45 percent. All six remaining studies (and the revised Clark University pilot study) showed a very consistent response in the narrow range of between 37.8 and 48 percent acceptance of reclaimed water for drinking. These studies also include more larger-scale samples than do those with higher acceptance rates. To date, therefore, the best estimate of public acceptance for drinking suggests that a majority of the public opposes it.

This generalization calls for immediate qualification. The acceptance rate may well rise in cities experiencing a water-supply crisis, in arid and semi-arid regions of the United States, and in communities that have had experience with water reclamation projects involving high-order uses of the water.

TABLE 7.4

*Public Acceptance of Renovated Wastewater
According to Potential Use*

| Use | Public Response | | |
| --- | --- | --- | --- |
| | Positive | Neutral | Negative |
| Drinking water | 48 | 8 | 44 |
| Cooking | 51 | 7 | 42 |
| Swimming | 78 | 7 | 15 |
| Washing clothes | 80 | 5 | 15 |
| Vegetable irrigation | 80 | 4 | 16 |
| Fishing | 81 | 6 | 13 |
| Industrial cooling | 94 | 3 | 3 |
| Golf course irrigation | 96 | 2 | 2 |

## Correlates of Attitudes: A Comparison of Findings

The record of research to date suggests that public response to using re-claimed wastewater relates to three factors—the adequacy of the present water supply and options for new sources, formal education, and knowledge or prior experience with water re-use. There is also evidence (less convinc-ing, however) that age, sex, occupation, and the quality of the present supply may enter into public assessments. Table 7.5 summarizes the results of studies on these relationships.

Perhaps the clearest factor influencing attitudes toward using reclaimed wastewater is one's *formal education*. Six of the seven studies which tested this relationship found it significant, frequently at the 0.01 significance level. Only the Ralph Stone and Company telephone survey (and the survey in Israel reported in Chapter 6) found no significant relationship, although the high refusal rate (31 percent) in the Ralph Stone and Company study may well account for this finding. In the Clark University study proper, 26 percent of the respondents with only a grade school education or less approved of reclaimed water for drinking, whereas 63 percent of those with some college education approved. In the Gallup Poll, 60 percent with only a grade school education or less indicated that they would object to drinking "recycled sew-age," whereas only 41 percent of those with a college education were simi-larly inclined. Amount of formal education, therefore, strongly colors attitudes toward water re-use.

Paralleling education is the *degree of knowledge or previous experience* that the respondent has had with water reclamation. Of the four studies which tested previous knowledge, the Johnson, Ralph Stone and Co., and Clark University study proper found a strong positive relationship (at 0.01), while the relationship was approaching significance in the Johns Hopkins

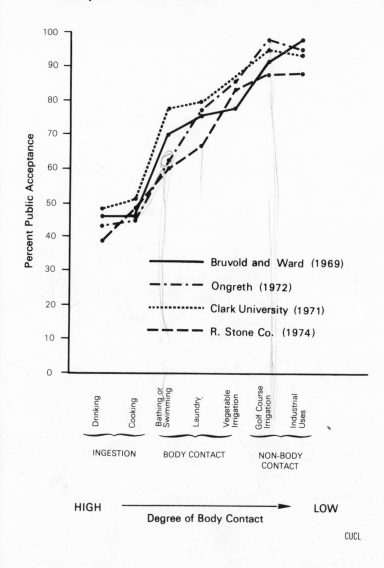

Figure 7.3: Attitudes toward Potential Uses of Reclaimed Water: A Comparison of
        Findings from Four Studies.

TABLE 7.5

*Public Acceptance of Renovated Wastewater:*
*A Comparative Summary of Findings from Eleven Studies*

| Study | Survey Characteristics | | | Willingness To Drink (%) |
| | Type of Survey | Scope; Location | Sample Size | |
| --- | --- | --- | --- | --- |
| Baumann (1965) | Telephone | National; Texas Kansas Mass. Ill. | 722 | 62.0 |
| Bruvold and Ward (1969) | Interview | Cal.; 2 Suburbs | 50 | 45.0 |
| Clark University Pilot Study (1970) | Interview | National; Gloucester Wilmington Kokomo Indianapolis | 220 | 29.0 (45.0) |
| Johns Hopkins University (1970) | Telephone | Baltimore County, Md. | 321 | |
| Pagorski (1971) | Interview | Park Forest, Ill. | 114 | 81.0 |
| Johnson (1971) | Interview | National; Phil. Camden Cincinnati Tucson Portland | 221 | 77.4 |
| Clark University Study Proper (1971) | Interview | National; Kokomo San Angelo Lubbock Santee Col. Springs | 420 | 48.0 |
| Bruvold and Ongreth; Bruvold and Ward (1972) | Interview | Cal.; 10 paired cities | 972 | 43.6 (may be (higher) |
| Carley (1972) | Interview | Denver | 447 | 38.7 (84.1) |
| Gallup Poll for AWWA (1973) | Mail Ques. | National | 2,927 (weighted) | 38.2 |
| R. Stone & Co. (1974) | Telephone | S. Cal. (10 cities) | 1,000 | 39.1 |

**Significant at 0.01     A.S. = approaching significance (0.10-0.
*Significant at 0.05     N.S. = not significant (> 0.10)

| | | | Correlates of Attitudes (Independent Variables) | | | | |
| Sex | Age | Income | Occupa-tion | Educa-tion | Knowledge of Re-use | Perceived Quality of Water | Cost |
|---|---|---|---|---|---|---|---|
| A.S. | N.S. | | A.S. | ★ | A.S. | | N.S. |
| | | | | | | ★ | |
| N.S. | N.S. | | | ★★ | ★★ | ★ (Negative Relationship) | |
| N.S. | ★★ | N.S. | N.S. | ★★ | ★★ | N.S. | N.S. |
| | | | | ★ | | | |
| ★ | ★ | | ? | ★ | | | |
| ★ | ★ | N.S. | ★ | ★★ | | | |
| | ★ | N.S. | N.S. | N.S. | ★ | | |

University survey. In the Clark University study, for example, of those respondents who could recall even having heard about the notion of water re-use, 52 percent were willing to drink reclaimed wastewater; only 24 percent of those uninformed about re-use were similarly inclined. Ralph Stone and Company found that the amount of "thought" given to the need for additional water resources was an important determinant of attitudes. The highest level of public acceptance (65 percent) uncovered in the Clark University five-city study occurred at Santee, where an extensive educational campaign and graduated experience with water reclamation had created an unusually knowledgeable public. Although Bruvold's ten-city study in California found no significant difference in attitudes toward drinking reclaimed water between matched communities possessing and lacking wastewater reclamation projects, he did find a clear tendency for users of water reclamation facilities to be less opposed than nonusers. So prior knowledge and experience also appear to be noteworthy factors in predicting acceptance levels.

There is some evidence that concern over the *quantity of water supply* increases acceptance of water re-use. Johnson found that relative approval of water re-use was much higher among consumers who did not perceive their water supply as adequate to meet peak demands over twenty years, while disapproval was greatest among those who saw the present supply as sufficient. Given the presence of supply options, Carley found only 37.8 percent acceptance of reclaimed water for drinking, but this rose to 84.1 percent once the options were removed. The Clark University study found that water-supply crises in Gloucester, Massachusetts, and San Angelo, Texas, had engendered more favorable community response to reclaiming wastewater than would normally be expected.

There are curiously divergent views on the role elicited by *perceived quality of the present supply*. Johnson contends that if an individual perceives the water he uses as polluted, he will be more willing to accept renovated water. Stone and Company argue differently: people with higher quality water generally have greater trust in the water treatment technology and responsible agencies and are, therefore, more positive toward accepting water re-use. The Clark University study tested this factor by asking for the respondent's evaluation of five characteristics of his drinking water but found no relationship, either positive or negative, between this evaluation and attitudes toward drinking reclaimed water. Clearly these alternative hypotheses need more research.

*Other social factors* play only weak or inconsistent roles in public acceptance of reclaimed water. There is fairly strong evidence in the Clark University, Carley, Gallup, and Ralph Stone and Company surveys that younger people are more inclined to drink renovated wastewater than are older persons, but the Johnson and Johns Hopkins studies found no significant relationship. There is also some evidence that women are more opposed to reclaimed water than men. Occupation and income appear to have little significance. Efforts by the Johns Hopkins group to relate political party affiliation with a respondent's attitude toward renovated wastewater proved unsuccessful. Likewise, the Johnson study showed no relationship between a respondent's attitude toward local government and his attitude toward using renovated wastewater.

Another factor possibly affecting public acceptance of renovated wastewater is the difference in *cost* of alternative sources of water supply. Perhaps increasing the price of water from conventional sources might result in greater public acceptance of a less costly renovated wastewater alternative. The Clark University study gave to three groups the choice of accepting a traditional source of supply, a reservoir, over renovated wastewater but at three different prices. One group was told that maintaining a conventional source of water supply would cost 25 percent more than utilizing renovated wastewater; the second group was told that the price would be 50 percent greater; and a third group was told that the average monthly bill would increase by 100 percent. The interviewer calculated that actual increase in the respondent's water bill and did not rely on the latter's interpretation of the meaning of the percentage increase. A fourth group was merely given the choice of utilizing reservoir water or renovated wastewater.

Changes in the price of water did not affect significantly a respondent's willingness to accept renovated wastewater. Increases in the price of water in order to maintain conventional sources of supply did not, in the Clark University study, have any statistically significant effect on public acceptance of renovated wastewater. A similar attempt to relate cost to attitudes in the Johns Hopkins study also failed to reveal any significant effect on acceptance levels.

Finally, it is important to assess the role that *psychological factors* may play in public response to water re-use. It is frequently asserted that there is a natural revulsion by the public to reclaimed wastewater. Clearly managers are concerned that they may launch ambitious educational programs only to find an irrational hostility based upon psychological aversion or avoidance. Noting that psychological repugnance is the reason most commonly cited by respondents for their unwillingness to accept reclaimed wastewater, Ralph Stone and Company sees greater difficulty in overcoming this aversion than simply education about the quality of the water.[25] Is there evidence to support a view that psychological repugnance is an important component of attitudes toward reclaimed water?

The only study to test this question in depth was a portion of the Clark University research designed by Duane Baumann and John Sims.[26] In the pilot study some twenty-nine sentence-completion items tested six psychological dimensions of attitudes: (1) a fear of, or disgust at, incorporation of or contact with the impure, (2) a faith in science and an accompanying trust in expertise versus suspicion of technology and mistrust of scientific authority, (3) an internal versus external locus of control, (4) a modern, innovative, progressive approach to problem-solving versus a traditional, conservative "holding on" to established methods, (5) a view of the world in which technology is seen as interfering with Nature's or God's ordained system, and (6) a deep essentially aesthetic commitment to the "natural" as opposed to what is perceived as artificial.

Unfortunately, the findings in the pilot study were largely negative—among people who varied widely in their acceptance of reclaimed water, there were

25. Ralph Stone and Company, p. II-36.

26. See Duane Baumann and John Sims, "Renovated Wastewater for Drinking: The Question of Public Acceptance," *Water Resources Research* 10 (1974), 675–682.

no statistically significant differences in the psychological dimensions. A revised psychological test employed in the study proper found that only two of ten sentence-completion items were significant. Both suggested that respondents favorable to drinking renovated wastewater tend to possess a greater confidence in the effectiveness of scientific technology in addressing environmental problems. But basically the research suggests an absence of any strong psychological basis for attitudes and argues that acceptance depends more upon what individuals know about re-use and their general educational level.

## The Prospect for Public Acceptance

The foregoing analysis of experience and survey research to date suggests the need for caution over the short run but for greater optimism over the long run. The prevailing levels of public acceptance of reclaimed water in American cities reveal widespread acceptance of lower-order (nonbodily contact) uses but majority opposition to ingestion uses. A ladder arrangement of attitudes toward reclaimed water suggests that public resistance is likely to increase as more intimate uses of the water occur. The existing state of knowledge concerning wastewater reclamation and re-use is still rudimentary—most urban Americans have not thought seriously about water re-use and are unfamiliar with the technology involved. Initial reactions to direct potable re-use are characteristically negative but improve with knowledge and experience. The present situation, in short, will allow a good deal of experimentation with lower-order uses of reclaimed water over the short run; meanwhile, individuals become more familiar with and knowledgeable about water reclamation, its technology, prospective uses, and the hazards discussed in Chapter 4.

The present state of public acceptance does not encourage premature or precipitous installation of direct potable re-use, however, except perhaps in cases of dire emergency. The highest levels of acceptance found in research thus far still leave limitless possibilities for community conflict. The present situation with regard to nuclear energy in the United States, in which two-to-one acceptance levels commonly prevail, indicates the potential for conflict inherent in premature deployment of a technology in which hazards are inadequately understood. More than majority support for high-order uses of reclaimed water will be required in American cities. Contrary to popular belief, power is widely dispersed in most American cities, with the proliferation of veto groups an inevitable result. Substantial consensus will likely be required to support high-order experimentation with reclaimed water. Given present knowledge and acceptance levels, such a consensus will require a lengthy process of education and experience.

Yet there are reasons to believe that a cautious deployment of water reclamation and re-use technology *over the long run* can meet with success. The evidence from survey research suggests that urban Americans are, on the whole, basically rational in their attitudes toward water re-use. Formal

education, experience with and previous knowledge about wastewater reclamation, and amount of thought given to water resources problems are the
factors that shape favorable public acceptance. The success of the elaborate educational campaign and experience with the ladder of water uses at
Santee in achieving widespread public support confirms the findings of survey research.

The state of the art of public-acceptance research supports, in short, the
innovation process advocated in this book. There is little need, given our
limited knowledge of health risks, to move prematurely to potable uses of
reclaimed water. Since the bulk of water use in American cities is nonpotable, there is widespread opportunity to accumulate experience with
water re-use for industry and agriculture, thereby releasing the present
supply for greater potable use. Such a process will permit the accumulation
of vitally needed experience with wastewater reclamation and re-use. It will
also permit a hesitant and uninformed public to gather knowledge, to test
its assumptions, to review successful experiments in the city and in nearby
communities, to allay fears. Such a gradual and cautious development of
water re-use offers a bright prospect for public acceptance.

But there is one important ingredient in the emergence of strong public
support which remains to be discussed. The research to date indicates that a
concert of support among water-supply managers and public health officials
is absolutely vital to widespread public acceptance. Division within the scientific and regulatory community over unresolved health issues could well lead
to substantial citizen opposition and rancorous community conflict. In Israel
the endorsement of the Ministry of Health was critical in public acceptance
(Chapter 6). The attitudes of managers become, therefore, a key factor for
acceptance and for the prospect for water re-use as a whole. The next several chapters inquire into the nature, bases, and significance of managerial
attitudes and behavior.

# Water Re-Use and the Professionals: Consulting Engineers and Public Health Officials

Thomas J. Nieman

**8**

In governmental organizations and private firms alike there is a trend toward increased professional specialization.[1] Since it extends into virtually all developmental areas, this change is not unique to any one aspect of planning and decision-making. To the extent that specialists or managers affect the development and utilization of innovations, it is important to determine the rationale for their recommendations and decisions.

For its part, the public often appears content to delegate decision-making power to specialists or managers by placing the problem outside its immediate sphere of concern. This abrogation of responsibility by the public is due at least in part to the increasing pressures of an urbanized environment.[2] The city offers too many potential contacts to be easily assimilated; in response, man orders his social relations so as to reduce diversity and to avoid excessive demands upon his time and emotions. W. R. Derrick Sewell, noting that "one of the most significant characteristics of modern society is its overwhelming dependence on experts," finds that professionals have the highest social status among experts.[3] Thus the way is open all too often for professionals to decide what the public wants and to direct the manner of obtaining these wants.

Of the professionals who guide the course of water development in the United States, consulting engineers and public health officials play a particularly central role. Engineers, for example, are the dominant force in the planning and development of water-supply and waste-disposal facilities.[4] Public health officials, on the other hand, are the key actors in safety policy for public water supply.[5] Consulting engineers are not likely to recommend a water supply alternative over the opposition of a department of public health. Likewise, public health officials are not likely to endorse a system whose safety and viability cannot be conclusively demonstrated by consulting engineers. Unless both groups agree as to the viability of any water supply alternative, the chance of its adoption (as is evident later in the book) is minimal. Therefore, knowledge of the position taken by these groups is a primary factor in the future acceptance of any water resource alternative.

In general, it appears that both groups tend to resist innovations, especially if they involve some risk. Carl Brunner contends:

> This is the responsibility that goes with their jobs. The consulting engineer is ethically, if not legally, responsible for the satisfactory operation of any system he designs. He cannot afford many mistakes and stay in business. While he may be convinced of the desirability of

1. Robert Perrucci, "Engineering, Professional Servant of Power," *American Behavioral Scientist*, 14 (1971) 492–506.

2. Stanley Milgram, "The Experience of Living in Cities," *Science*, 167 (March 13, 1970), 1461–68.

3. W. R. Derrick Sewell, "Environmental Perceptions and Attitudes of Engineers and Public Health Officials," *Environment and Behavior*, 3 (1971), 23–59.

4. Gilbert F. White, "Man's Alteration of the Water Cycle," in *Attitudes toward Water: An Interdisciplinary Exploration*, ed. F. L. Strodtbeck and G. F. White (privately distributed, 1970), pp. 1–4; Neil L. Ackerman, "Community Attitudes toward Recycling Wastewater," unpublished Master's thesis, Department of Geography, Southern Illinois University (1970), p. 8; Frank M. Middleton, "Municipal Drinking Water and Waste Water—A National Problem," paper presented at a seminar on Water and Sewage Problems of Western New York, Rochester, New York (May 23, 1963).

5. Middleton, p. 9.

6. Letter from Carl A. Brunner, Advanced Waste Treatment Research Laboratory, Environmental Protection Agency, Cincinnati, Ohio (January 7, 1973).

7. Sewell, p. 53.

8. Gilbert F. White, "Formation and Role of Public Attitudes," in *Environmental Quality in a Growing Economy*, ed. Henry Jarrett (Baltimore, Johns Hopkins Press, 1966), pp. 105–127.

9. Sewell, p. 40.

10. White, "Man's Alteration of the Water Cycle."

11. Harvey S. Perloff and Lowdon Wingo, Jr., eds., *Issues in Urban Economics* (Baltimore, Johns Hopkins Press, 1968), pp. 12–14.

12. White, "Formation and Role of Public Attitudes," p. 125; W. R. Derrick Sewell, "The Role of Attitudes of Engineers in Water Management," in *Attitudes toward Water*, p. 64/4.

13. W. R. Derrick Sewell and Brian R. Little, "Specialists, Laymen and the Process of Environmental Appraisal," *Regional Studies*, 7, No. 2 (1973), 161–171.

wastewater reuse and may be quite liberal in his general attitude, he will resist innovation if there is much risk of failure. The public health official has a strong responsibility to protect the public. In the case of potable water, he falls back on empirical results of many years of water treatment experience to judge what is safe.[6]

Sewell also notes that these professional groups betray all the characteristics of a closed system:

> Their views seem to be highly conditioned by training, adherence to standards and practices of the respective professions, and allegiance to the agency's or firm's goals or mission. Both groups believe they are highly qualified to do their respective jobs and that they act in the public interest. Contact with representatives of other agencies or the general public, however, is considered either unnecessary or potentially harmful. There appears to be general satisfaction with past policies and practices, and few, if any, major alterations are suggested.[7]

Decision-making by professionals, however, does not rely solely on the information provided by the professional organization. It is also a product of individual goals and beliefs which generally enter into the decision-making process in three forms: (1) the personal attitude of the decision-maker, (2) his perception of what others prefer, and (3) his opinion as to what others should prefer.[8] Though personal attitudes play a major role in decision-making, professionals remain reluctant to deviate from established procedure and what they perceive as acceptable by the public.[9] Gilbert F. White, for example, suggests that "suppliers of domestic water do not consider using polluted sources if other sources are available because they believe the consumer would object."[10] Concerns beyond cost of supply and treatment, in short, can be major factors in decisions.

There is, however, no direct mechanism to inform managers about what people actually want. The manager must perforce make decisions based on his perception of what the public ought to want. The most influential factor, then, in the course of daily environmental decision-making is an assessment of public attitudes by a host of public officials who rely on perception rather than fact.[11]

In the case of municipal water supply, White sees a strong tendency to rely on an external authority for judgment. Moreover, this judgment itself is based on the professional's perception of what people will accept and/or support. As a result, a small group of professionals, on the basis of unconfirmed perceptions of what the people want, guide the course of water development in the United States.[12]

In their capacity as professionals, consulting engineers and public health officials are similar. As technical advisors they affect policy decisions which, in turn, play an instrumental role in the adoption of municipal water supply and treatment plans. Sewell argues that these professionals claim to portray only their own views and that they see public involvement as a hindrance to effective action.[13] Many professionals contend, on the other hand, that they can adequately represent the views of the public. In the choice of a public water-supply alternative in the Grand River Basin Region, Canada, however, Ian McIver found that water-supply officials overwhelmingly cited the river as the choice of the public, but the public, when asked, did not agree.[14] Clearly, information gaps exist between officials and the public as well as among the officials themselves. Their perception of the attitudes of others are either genuine misinterpretations or really projections of their own biases assimilated into the institutional decision-making framework.

## Direct Potable Re-Use: A Last Resort

This chapter proposes to assess the attitudes of consulting engineers and public health officials toward the adoption of renovated wastewater for potable use and to discuss behavioral bases of these attitudes. As managers of a diminishing high-quality supply of raw water, their negative attitudes to the use of lower-quality water sources bear heavily on the future of adequate community supplies. The policy statements of the American Water Works Association and the Water Pollution Control Federation cited in Chapter 1 illustrate the serious reservations among professionals over direct re-use for potable purposes. The Water Reuse Committee of the Water Pollution Control Federation specifically asks that potable re-use be advocated only when "all possibilities of conserving water, redistributing population, providing dual systems, or other techniques acceptable to the population requiring the water have been exhausted."[15] John Parkhurst, one of the pioneering managers of re-use, states that "experience to date indicates that water supply agencies and people in general will look to any available existing natural supply which they can afford prior to investing in reusable water."[16] While confident of the technology available, Russell Culp nonetheless argues that "no reason now exists for the direct reuse of reclaimed water for potable supply . . ."[17] The very idea, in fact, "tends to shock those in water management."[18]

It is apparent that the adoption of renovated wastewater for potable use will be considered as a resource alternative only as a last resort, and clearly there is professional concern, as there should be, over public health and economic issues. But it would be instructive to know if there are other roots to this resistance to re-use options as well. Water supply managers and professionals tend to assume that the public will be unalterably opposed to renovated wastewater for potable use. In light of the findings in Chapter 7, however, it appears that this perception may be a misinterpretation of public attitudes. Feldman shows in Chapter 6 that favorable public response was forthcoming in Israel. Apparently education, past experience, and severity of need all play a significant role in public response. Kasperson in Chapter 7 proposes that graduated experience with a hierarchy of uses constitutes a mechanism for gaining public acceptance. But he also makes clear that the initial burden of acceptance falls directly on the water manager, who must make the first positive step toward adoption. Managers, as the next chapter will show, occupy a very strategic position in initiating the innovation. Their

14. Ian McIver, "Municipal Water Supply in Grand River Basin Region," in *Perceptions and Attitudes in Resources Management*, ed. W. R. Derrick Sewell and Ian Burton (Ottawa, Canada, Policy Research and Coordination Branch, Dept. of Energy, Mines, and Resources, 1971), p. 58.

15. Water Pollution Control Federation, Water Reuse Committee, "Policy Recommendation on Direct Reuse of Water" (Washington, 1971) (Mimeographed).

16. John D. Parkhurst, "The Feasibility of Water Reclamation (Renovation and Reuse)," testimony before the State Water Resource Control Board, San Jose, California, December 11, 1972. (Mimeographed.)

17. Russell Culp, "Uses of Reclaimed Water," *Water and Wastes Engineering*, 6 (1969), 54.

18. Ralph E. Fuhrman, "Adaption of Known Principles and Techniques of Waste Water Management to Specific Environmental Situations and Geographical Conditions," *Journal of the Water Pollution Federation*, 68 (1969), 619–621.

support was crucial for public acceptance in Israel (see Chapter 6). This chapter explores the rationale and bases for the attitudes of consulting engineers and public health officials. Hopefully, this discussion will illuminate somewhat the nature and strength of professional support for and resistance to renovated wastewater for potable uses.

## Data Collection and Sample

To obtain information concerning the future acceptance of renovated wastewater for potable use, researchers examined the attitudes and perceptions of consulting engineers and public health officials. The specific aims were: (1) to determine the extent to which the attitudes of consulting engineers and public health officials act as a hindrance to the future development of planned potable re-use systems, and (2) to explain these attitudes in terms of specific sociological, professional, and personal constructs.

Data relevant to the determination of the attitudes and perceptions of these two groups include the overt expression of an attitude toward re-use, projection of an attitude to other individuals, and the perception of the attitude of other professionals. Interviews also sought information concerning social background, professional education, and actual professional practice. Finally, the research included instruments to examine such psychological constructs as fate control, a progressive versus traditional stance to innovation, and revulsion toward body waste.

After formulating and pretesting the study instrument, researchers drew the sample (Table 8.1) of ninety-eight consulting engineers from three regions of the United States (Figure 1.1), which included the largest consulting firms—as described by the *American Institute of Consulting Engineers Directory*. The sample includes only those consulting engineering firms actively engaged in the development of municipal water supply and treatment facilities. Researchers then drew the sample of twenty-two public health officials from the state agencies in the same regions.

## Professional Attitudes toward Potable Re-Use

Whereas it is clear that professionals evince a conservative attitude toward the potable use of renovated wastewater, previous studies have not delineated the characteristics of and reasons for this position. The overtly expressed attitudes of consulting engineers in this study suggest that the statements expressed in the professional journals do represent the practitioners as a whole (Table 8.2). Only one in five consulting engineers unqualifiedly endorses direct potable re-use, and the reasons tend to divide evenly among economic, operational, and technological considerations. As for public health officials, the majority hesitates to accept renovated wastewater because of skepticism primarily over operations (36.4 percent) or technology

TABLE 8.1

*Location and Number of Interviews*

TOTALS

Consulting engineers: 33 firms, 98 interviews

Public Health officials: 10 state departments, 22 interviews

| State | No. of Firms | No. of Interviews |
|---|---|---|
| EAST | | |
| 12 Engineering firms (33 interviews) | | |
| New York | 4 | 16 |
| Massachusetts | 3 | 7 |
| Maryland | 2 | 7 |
| Connecticut | 1 | 2 |
| Washington, D.C. | 1 | 1 |
| 4 State health departments (7 interviews) | | |
| New York | | 2 |
| Massachusetts | | 2 |
| Maryland | | 1 |
| Connecticut | | 2 |
| MIDWEST | | |
| 9 Engineering firms (33 interviews) | | |
| Ohio | 3 | 8 |
| Illinois | 3 | 13 |
| Missouri | 2 | 7 |
| Iowa | 1 | 5 |
| 3 State health departments (9 interviews) | | |
| Ohio | | 3 |
| Illinois | | 3 |
| Missouri | | 3 |
| WEST | | |
| 12 Engineering firms (32 interviews) | | |
| Texas | 2 | 6 |
| Arizona | 2 | 4 |
| California: | | |
| Los Angeles | 3 | 8 |
| San Francisco | 5 | 14 |
| 3 State health departments (6 interviews) | | |
| Texas | | 2 |
| Arizona | | 2 |
| California | | 2 |

(40.9 percent). Only one in twenty wholeheartedly endorses this innovation. For the most part their concerns center on possible human error in plant operations and the safety factor involved in the process itself.

To shed further light on the nature of the attitudes of the two groups toward

TABLE 8.2

*Overtly Expressed Attitudes toward Direct Potable Re-Use*

| Personal Feelings Expressed | Consulting Engineers | | Public Health Officials | |
|---|---|---|---|---|
| | No. | % | No. | % |
| For recycling: no qualifications | 20 | 20.4 | 1 | 4.5 |
| Hesitant to accept: economics | 20 | 20.4 | 3 | 13.6 |
| Hesitant to accept: operations | 21 | 21.4 | 8 | 36.4 |
| Hesitant to accept: technology | 21 | 21.4 | 9 | 40.9 |
| Hesitant to accept or reject: Aesthetics | 16 | 16.3 | 1 | 4.5 |

$X^2 = 9.16$     Sig. at .10     N = 120

the utilization of renovated wastewater in a municipal situation, the study utilized a projective psychological technique in the form of a modified Thematic Apperception Test (TAT) (see Appendix 2).[19] This test requires the respondent to write a brief story interpreting a photograph (Figure 8.1) which depicts a community group in the process of solving an impending water

Figure 8.1: Photograph Used in the Thematic Apperception Test.

TABLE 8.3

*Projection of Attitude toward Direct Potable Re-Use*

| Attitude Projected | Consulting Engineers | | Public Health Officials | |
|---|---|---|---|---|
| | No. | % | No. | % |
| For recycling: no qualifications | 23 | 23.5 | 1 | 4.5 |
| For partial use and other uses | 15 | 15.3 | 3 | 13.6 |
| Uncertain and postpone decision | 23 | 23.4 | 3 | 13.6 |
| Use only as a last resort | 19 | 19.4 | 8 | 36.4 |
| Reject: no qualifications | 18 | 18.4 | 7 | 31.8 |

$X^2 = 7.91$      Sig. at. 10      N = 120

shortage. TAT theory assumes that the respondents unconsciously project their own attitudes onto the group in the picture without directly stating to the interviewer their positions on the acceptance of renovated wastewater for potable use.

In their respective solutions to the impending water shortage, consulting engineers and public health officials projected attitudes consistent with the overt expression of their feelings (Table 8.2). Of the consulting engineers, 23.5 percent openly accept renovated wastewater as a solution to the problem, while 18.4 percent reject it out of hand (Table 8.3). In general, caution pervades their approach. Public health officials, on the other hand, project a definite reluctance to accept recycling—31.8 percent oppose it as a solution to the water shortage problem, and 36.4 percent accept it only as a last resort. Only one in twenty support it without reservation. These figures are remarkably consistent with the overt expressions of attitudes and provide a validity check for the research.

A further indication of the acceptability of an innovation is the length of time it requires for societal adoption. One indication of a person's attitude toward an innovation, therefore, is the perception of the time required for societal acceptance. Consulting engineers and public health officials differed in their estimates of the time lapse which might precede adoption of renovated wastewater as a viable alternative for potable use in the United States (Table 8.4). Whereas one in four consulting engineers predicted a time lapse of ten years or less, only one in twenty public health officials shared this view. By contrast, one fifth of the engineers but over half of the public health officials forecast a minimum of thirty years and/or possibly never. This contrast is consistent with the stance of general caution expressed by the consulting engineers and the more resistant attitude of the public health officials.

Both groups see their own attitudes as consistent with prevailing views in their professions (Table 8.5). Consulting engineers perceive the attitude of other consulting engineers as generally positive but initially cautious (64.4 percent within the individual's own organization and 53.3 percent within the profession). Public health officials, on the other hand, perceive the attitude of

TABLE 8.4

*Perceived Time Required for Adoption of Direct Potable Re-Use in the United States*

| Time Projected | Consulting Engineers | | Public Health Officials | |
|---|---|---|---|---|
| | No. | % | No. | % |
| Less than 10 years | 25 | 25.5 | 1 | 4.5 |
| 10–19 years | 29 | 29.5 | 5 | 11.7 |
| 20–29 years | 23 | 23.5 | 4 | 18.2 |
| Over 30 years to never | 21 | 21.4 | 12 | 54.5 |

$X^2 = 13.03$     Sig. at .01     N = 120

other public health officials as basically negative (40 percent within the individual's own organization and 50.5 percent within the profession).

The data results confirm that both groups are conservative in their attitudes toward potable use of renovated wastewater. Public health officials, who see such systems as possible only in the far distant future, believe that other health officials share this view. They perceive consulting engineers as more likely to accept the process for the sake of economic gain. In the same vein, consulting engineers, while cautious on the issue of potable use of renovated wastewater, are not adamantly opposed to it. Their attitude tends toward the ambivalent. Given time, they visualize its use as a routine occurrence. At present, however, they choose not to risk an innovation that might possibly prove unsuccessful. They perceive members of their own profes-

TABLE 8.5

*Perception of the Attitudes of Other Consulting Engineers and Public Health Officials*

| | Consulting Engineers | | Public Health Officials | |
|---|---|---|---|---|
| | No. | % | No. | % |
| *Perceived Attitude of Others Within the Organization*[a] | | | | |
| Accept recycling to cautious | 58 | 64.4 | 8 | 40.0 |
| Attitude varies | 13 | 14.4 | 2 | 10.0 |
| Reject to accept as last resort | 19 | 21.1 | 10 | 50.0 |
| *Perceived Attitude of Others within the Profession*[b] | | | | |
| Accept recycling to cautious | 48 | 53.3 | 12 | 50.5 |
| Attitude varies | 20 | 22.2 | 1 | 4.0 |
| Reject to accept as last resort | 22 | 24.4 | 10 | 45.5 |

[a]$X^2 = 8.68$     Sig. at .05     N = 120
[b]$X^2 = 8.23$     Sig. at .05     N = 112

sion as expressing similar views and public health officials as unduly conservative.

## Whither the Public and the Politicians?

Consulting engineers and public health officials are apparently convinced that the general public is unalterably opposed to the potable use of renovated wastewater. H. Christopher Medbery, for example, while proposing utmost consideration for re-use for potable supply, states flatly that "the public is not ready to accept it directly for drinking water purposes."[20]

The New York City water crisis during the 1949–1950 drought in the Northeast suggests the importance of professional perceptions of public attitudes.[21] The immediate response in New York was to construct a 100-mgd pumping plant to draw water from the Hudson River, a polluted raw source. After much controversy among the water-supply factions of the city over the quality of the raw water, however, officials decided not to utilize the plant. In fact, after the crisis had passed, they dismantled it altogether.

The chief engineer of the Department of Water Supply, Gas, and Electricity for the City of New York opposed the proposal to utilize Hudson River water because, "New York public sentiment would be against drinking from the Hudson River."[22] A more recent investigation supports this claim and expands upon alleged public rejection by proposing that psychological obstacles are not unique to New York City, that many United States cities have serious aversions to using river water as a source of potable supply.[23] Contending that there was no formal attempt to determine public sentiment, Neil Ackerman dismisses the assumption of public rejection of Hudson River water.[24] Finding only evidence of conjecture on the part of city decision-makers, he concludes that the public was a passive factor in the controversy. Since the issue did not require a referendum, water managers (that is, the political and professional members of the Board of Water Supply), and not the public, made the decision.

Did the public react negatively to the use of the Hudson River water? The question was partially answered in 1968 when, after another water crisis, the city of New York built and utilized a new pumping station on the Hudson River. The news media emphasized the polluted state of the river; yet during the crisis period, the public accepted the 21 billion gallons of river water provided by the city.[25]

There is some reason to believe that the aversion to re-use of wastewater may be lessening over time. The use of sewage effluent as a component of the water supply of 145 of 155 United States cities surveyed by the U.S. Public Health Service,[26] the favorable public reaction (as noted in Chapter 7) to the renovated wastewater recreational facilities at Lake Tahoe and Santee, California, the success of re-use in Israel (Chapter 6), and the well publicized notion of recycled wastewater in space flights—all could act to reduce public resistance.[27] Increasing public awareness of successful adoptions of renovated wastewater and the advantages of that use could serve to assure

20.  H. Christopher Medbery, "Preparing for Tomorrow's Needs," *Journal of the American Water Works Association*, 58 (1966), 930–938.

21.  Jack Hirshleifer, James C. DeHaven, and Jerome W. Milliman, *Water Supply: Economics, Technology and Policy*, rev. ed. (Chicago, University of Chicago Press, 1969), pp. 255–288.

22.  Editorial, *New York Times* (December 18, 1949).

23.  John T. Kearns, "Water Conservation and Its Application to New England," *Journal of the American Water Works Association*, 58 (1966), 1379–84.

24.  Ackerman, p. 22.

25.  Dwight F. Metzler and Heinz B. Russelmann, "Wastewater Reclamation as a Water Resource," *Journal of the American Water Works Association*, 60 (1968), 95–102.

26.  Louis Koenig, *Studies Relating to Market Projections for Advanced Waste Treatment*, U.S. Department of Interior, Publication WP-20-AWTR-17 (Washington, D.C., 1966).

27.  Frank P. Sebastian, "Wastewater Reclamation and Reuse," *Water and Wastes Engineering*, 7 (1970), 46–47.

**Thomas J. Nieman**

28. Metzler and Rus-
selmann, p. 101.

TABLE 8.6

*Perception of Public Attitudes toward Renovated Wastewater*

| Perceived Attitude of the Public | Consulting Engineers | | Public Health Officials | |
|---|---|---|---|---|
| | No. | % | No. | % |
| First reject—then accept | 53 | 55.1 | 4 | 18.2 |
| Hesitant: safety and aesthetics | 33 | 33.7 | 11 | 50.0 |
| Accept as last resort, reject | 11 | 11.2 | 7 | 31.8 |

$X^2 = 12.25$     Sig. at .02     N = 120

routine acceptance in the future. Though reluctance on the part of the consumer is understandable, the obstacle is, as the preceding chapter makes clear, not insurmountable. Two proponents of water re-use have suggested that "the challenge of acceptance is greatest with the water utility managers."[28]

Interviews with consulting engineers and public health officials in this study confirm this propensity to assume public resistance. Over half of the consulting engineers believe that the public will initially reject but ultimately accept the idea in time and after education, but fewer than one in five public health officials share this belief (Table 8.6). In fact, half predict that the public will reject renovated wastewater on the basis of safety and/or aesthetics. Only one in three consulting engineers was equally pessimistic.

Both consulting engineers and public health officials anticipate some degree of opposition from political figures as well as from the general public. Nearly twice as many public health officials as consulting engineers, however, see the attitudes of political officials as a primary, or at least secondary, obstacle to the adoption of renovated wastewater for potable use. Nearly

TABLE 8.7

*Perception of Opposition by Politicians*

| | Consulting Engineers | | Public Health Officials | |
|---|---|---|---|---|
| | No. | % | No. | % |
| *Expressed Political Obstacle*[a] | | | | |
| Primary obstacle | 12 | 12.1 | 7 | 31.8 |
| Secondary obstacle | 28 | 28.6 | 10 | 45.5 |
| Not a factor | 58 | 59.2 | 5 | 22.7 |
| *Projected Political Concern*[b] | | | | |
| Concern expressed | 12 | 12.1 | 10 | 45.5 |
| Concern not expressed | 86 | 87.8 | 12 | 54.5 |

[a]$X^2 = 10.51$     Sig. at .01     N = 120
[b]$X^2 = 11.11$     Sig. at .001     N = 120

TABLE 8.8

*Perception of Opposition by the General Public*

| | Consulting Engineers | | Public Health Officials | |
|---|---|---|---|---|
| | No. | % | No. | % |
| *Expressed Public Obstacle*[a] | | | | |
| Primary obstacle | 48 | 49.0 | 17 | 77.3 |
| Secondary obstacle | 20 | 22.2 | 2 | 9.1 |
| Not a factor | 30 | 30.6 | 3 | 13.6 |
| *Projected Public Concern*[b] | | | | |
| Concern expressed | 27 | 27.6 | 12 | 54.5 |
| Concern not expressed | 71 | 72.4 | 10 | 45.5 |

[a]$X^2 = 5.80$     Sig. at .10     $N = 120$
[b]$X^2 = 4.80$     Sig. at .05     $N = 120$

four times as many public health officials as consulting engineers indicate concern for political implications (Table 8.7).

When it comes to possible opposition by the general public, this attitudinal pattern remains consistent. Once again, more public health officials than consulting engineers see public opposition as a serious obstacle. This pattern holds up both in the opinions expressed in formal interviews and in those projected in the TAT Test (Table 8.8). Since the status of public health officials hinges on public trust and involves heavy responsibilities, the roots of this greater concern are obvious. Over the years they have become respected as protectors of the public welfare and are reluctant to risk their reputations, particularly in the face of professional misgivings over a controversial issue such as the potable use of recycled wastewater.

Finally, the two groups differ in other concerns over potable re-use (Table 8.9). Public health officials, for example, express more concern over the adequacy of technological development. In both the formal interviews and the TAT tests, twice as many public health officials as consulting engineers see technology and health issues as important obstacles. But not surprisingly, consulting engineers are much more concerned with economic feasibility.

## The Contrast in Attitudes

The attitudes of consulting engineers and public health officials discovered in this research are consistent with previous suggestions. Both groups are conservative in their attitudes and perceptions concerning the potable use of renovated wastewater. Neither group entertains the possibility of outright acceptance. Rather, their feelings range from "possible but cautious acceptance" by consulting engineers to "last-resort acceptance" or "no acceptance" by public health officials.

**Thomas J. Nieman**

TABLE 8.9

*Other Concerns*

|  | Consulting Engineers | | Public Health Officials | |
|---|---|---|---|---|
|  | No. | % | No. | % |
| Technology: Expressed Obstacle[a] | | | | |
| Primary obstacle | 12 | 12.2 | 6 | 27.3 |
| Secondary obstacle | 19 | 19.4 | 7 | 31.8 |
| Not a factor | 67 | 68.4 | 9 | 40.9 |
| Technology (Health Safety): Projected Concern[b] | | | | |
| Concern expressed | 22 | 22.4 | 10 | 45.5 |
| Concern not expressed | 76 | 77.6 | 12 | 54.5 |
| Economics: Expressed Obstacles[c] | | | | |
| Primary obstacle | 34 | 34.7 | 7 | 18.2 |
| Secondary obstacle | 27 | 27.6 | 4 | 18.2 |
| Not a factor | 37 | 37.8 | 14 | 63.6 |

[a]$X^2 = 6.13$    Sig. at .05    $N = 120$
[b]$X^2 = 3.76$    Sig. at .10    $N = 120$
[c]$X^2 = 4.99$    Sig. at .10    $N = 120$

Both groups appear willing to express their feelings in response to direct questioning. Also, they tend to project these feelings onto other members of their profession. Public health officials, for example, perceive the implementation of renovated wastewater as a remote possibility and view other health officials as holding similar attitudes. They see consulting engineers as more likely to accept the process for the sake of economic gain. In their capacity as public servants, they are justifiably concerned with the health issues and are sensitive to the possible political ramifications of an innovation that might face a hostile political officialdom and general public. Seeking support for their own negative position, they then tend to project these feelings onto the public.

Consulting engineers are not amenable to the adoption of renovated wastewater, but they are not adamantly opposed to it either. They are cautious. Given time they visualize the adoption of water re-use systems, even for high-order uses, as a routine occurrence. At the present time, however, they choose not to risk an innovation that may possibly lead to problems. They see members of their own profession as holding the same views, but believe that public health officials are unduly conservative. More confident of re-use themselves, they are less concerned over adverse political reaction and public opinion than public health officials. The public, in their view, may initially reject the concept of direct potable re-use, but will eventually accept it as its merits become known. Unlike public health officials, they do not view technology as a serious problem; rather, they are concerned with economic feasibility. While consulting engineers may appear more amenable to the adoption of renovated wastewater for potable use than public health officials, they are still far from pioneers in the adoption process. More time and persuasion will be required to convince both groups of the potential of renovated wastewater. There are clear indications, however, that the public health professionals constitute a serious source of opposition to high-order uses of renovated wastewater, at least at this stage of technological development and scientific understanding.

## Sociological Bases

A generally accepted proposition of sociology is that group membership influences an individual's perception, interpretation, and reaction to society and the environment. By means of socialization, the individual acquires and develops attitudes.[29] According to this theory, the attitudes of people are basically a function of social origins, economic class, group membership (age, sex, religion, political position, and occupation), and educational status.[30] Edward Suchman hypothesizes that a combination of cultural values, social pressures, and individual needs underlies all health attitudes and related subsequent behavior.[31] The specific areas of the socialization process examined in this study fall into four major categories: (1) social origins and background of the individual prior to professional training, (2) professional training, (3) professional socialization, and (4) regional differences. Examining the social development of public health and consulting engineering professionals offers some insights into the nature of the attitudes profiled above.

*Social origins and background*. In developing techniques for measuring attitudes toward fluoridation, C. Michael York found that age, education, sex, and socioeconomic status were all useful attitudinal correlates.[32] This work suggested that older, less-educated, females of lower economic status were the most likely to oppose the use of fluoridation in the drinking water. Chapter 5 suggests that early adopters of re-use most likely will be high-status communities which are populated by young, educated individuals who have previous experience with or knowledge of wastewater reuse. Yet another study of engineers concluded that social origins have little significance— except for possible contact with professionals—for career and attitudinal attainment beyond the initial critical decision to seek an engineering education.[33]

Social origins and backgrounds may act to decrease the variance among attitudes that exists within a profession and, conversely, increase the difference among professions. Relationships among psychological, sociological, and biological characteristics may account for the similarities in individuals who enter a specific profession. Anne Roe described these characteristics as occurring in a regular pattern of frequencies such that individuals displaying a configuration of specific characteristics may be suited to certain professions.[34] Thus if the two groups in the present study display distinct social

29. *Socialization*: "simultaneously describes a process, or import, external to the person, the individual's experience of the process, and the end product or output. [It] refers to the process whereby individuals acquire the personal system properties—the knowledge, skills, attributes, values, needs and motivation, cognitive patterns—which shape their adaption to the physical and socio-cultural settings in which they live." Alex Inkeles, "Social Structure and Socialization," in *Handbook of Socialization and Research*, ed. D. A. Goslin (Chicago, Rand McNally, 1969), pp. 615–632.

30. A. L. Knutson, *The Individual, Society, and Health Behavior* (New York, Russell Sage, 1965), p. 137.

31. Edward A. Suchman, "Health Attitudes and Behavior," *Archives of Environmental Health*, 20 (1970), 105–110.

32. C. Michael York, "Opinions Relating to Fluoridation: Development of Measures," *Water Resources Bulletin*, 6 (1971), 920–924.

33. Joel E. Gerstl, and S. P. Hutton, *Engineers: The Anatomy of a Profession* (London, Tavistock, 1966), p. 27.

34. Anne Roe, *The Psychology of Occupations* (New York, Wiley, 1956), pp. 103–115.

35. This group of people appears to follow a view similar to that of Max Weber's Protestant ethic. According to Weber, man should be individualistically oriented, want to work hard and get ahead, and want to make money. William F. Whyte, *Money and Motivation* (New York, Harper, 1955), pp. 12–13.

36. Robert Eichhorn, "The Student Engineer, in *The Engineer and the Social System*, ed. Robert Perrucci and Joel Gerstl (New York, Wiley, 1969), p. 142.

37. Howard S. Becker et al., *Boys in White* (Chicago: University of Chicago Press, 1961); Harold Leif and Renée C. Fox, eds., *The Psychological Basis for Medical Practice* (New York, Harper and Row, 1963), pp. 12–25.

38. Elmer L. Struening and Stanley Lehmann, "Authoritarian and Prejudicial Attitudes of University Faculty Members," in *The Engineer and the Social System*, p. 162.

39. Ibid. (attributed to Robert Eichhorn).

40. Daniel J. Levinson, "Medical Education and the Theory of Adult Socialization," *Journal of Health and Social Behavior*, 7 (1967), 253–265.

41. M. Alfred Haynes, "Professionals and the Community," *American Journal of Public Health*, 60 (1970), 519–523.

42. Marshall W. Raffel, "Education for Health Services Administration," *American Journal of Public Health*, 60 (1970) 982–987.

backgrounds, a difference in attitude toward water re-use could well result. That is, if such variables as place of childhood, parental occupation, education, and religion differ between groups, they may explain differences in attitude toward renovated wastewater.

The findings on the social origins and background of consulting engineers and public health officials, however, reveal similarities rather than major differences. Members of both professions tend to come from politically conservative, Protestant[35] families of limited economic means; parental education does not surpass the high-school level, and the father's occupation is non-professional. These data are consistent with the previously developed premise that both consulting engineers and public health officials come from modest socioeconomic origins.[36] The two groups also resemble one another in social background. Inclined to remain within the geographic locale where they were born and reared, they prefer to modify, in a slightly liberal direction, the religious and political positions of their parents. In brief, consulting engineers and public health officials are little different from their parents. Their backgrounds make for a generally conservative approach to social innovation and change.

*Professional training*. Professional training is usually a major factor in professional socialization.[37] The student, influenced by the educational process, develops and shapes a belief system similar to those he emulates, thereby establishing a professional identity. This process explains why professionals appear to be more similar to each other than to other professional groups.

Professional education appears to set apart consulting engineers and public health officials from other professions. The engineer's education, which avoids the social sciences and humanities, forces him to draw on his own social background and that of his instructors—who are primarily engineers—for his concept of man and society.[38] This is consistent with the description of the engineer as a "narrowly focused, vocationally oriented, and socially conservative individual who enters his studies as a speedy road to professional, economic and social status."[39]

This description also fits public health officials with engineering backgrounds. Those who have a medical background (medical doctors) follow a paradigm of adjustment similar to that of engineers, but to a greater extent. In the medical school experience, the student is discouraged from coming into contact with extracurricular influences.[40] This process may contribute to a paternalistic approach (an assumption that they always know best) among public health officials.[41] This attitude may also become self-perpetuating, because public health officials, like other professionals, tend to hire people nearly identical to themselves.[42]

There is general agreement among social scientists that professional training plays a major role in professional socialization. In this respect the data indicate that consulting engineers and public health officials are very similar. Both groups attended approximately the same schools and attained comparable levels of educational achievement. Thus, from an educational point of view, it is difficult to explain any major attitudinal variation. The fact remains, however, that upon graduation, some individuals with engineering degrees choose to enter the field of public health rather than consulting practice. The extent to which this has affected their attitude toward re-use unfortunately is unknown.

*Professional socialization*. It is conceivable that after professional training, actual practice in the profession may contribute most to the homogeneity within various professions. As the profession (authority figure) socializes the young professional, he, in turn, socializes the profession. This may cause the profession to change its expectations to conform to the young professional's report of revised norms applied to new situations. This reciprocal relationship between the authority figure and the young professional may play a large part in the development of general similarities within the profession.[43]

In addition, repetitive actions tend to reinforce past experience and knowledge. The successful use of a particular problem-solving methodology will gradually be accepted and perpetuated as the proper solution to certain problems. The nature of the professional organization with its code of ethics and its professional standards and the individual's need to conform to established behavioral patterns further reinforce accepted methods and solutions.[44] This coincides with the tendency among professional resource managers to seek support for their decisions from their colleagues while ignoring or insulating themselves from the public.[45] The profession, as a socializing agent, contributes in this way to the attitudes that consulting engineers and public health officials hold toward potable re-use. In the sense that professions represent closed or closely controlled organizations in which entrance can be gained only through the consent of the gatekeepers, it is likely that social differences will diminish.

Other factors which may account for attitudinal differences include age, time spent in the profession, tenure with the present employer, and income level. As noted in Table 8.10, the mean age of public health officials is fourteen years greater than that of consulting engineers. While engineers are approximately evenly distributed among all age groups, 64 percent of the public health officials are over the age of fifty. This disparity of age corresponds with the fact that health officials have more professional experience. They have been in the profession an average of five years longer than the consulting engineers. Similarly, public health officials have been with the present employer an average of 13.5 years compared to a mean of nine years for consulting engineers.

Despite this greater age, professional experience, and tenure with their present employer, consulting engineers have, in a briefer time span, achieved a higher salary level. Interestingly, only 36 percent of the consulting engineers receive salaries in the $15,000–$25,000 range, which is the level of 63 percent of the public health officials. The salary distribution of consulting engineers reveals greater spread and variance—there are some very high salaries of over $50,000 and some lower salaries of less than $10,000. While a consulting engineer may begin his career at a lower salary than the public health official, opportunity exists to rise rapidly in salary level, a situation not available to public health officials.

43. Matilda W. Riley et al., "Socialization for the Middle and Later Years," in *Handbook of Socialization Theory and Research*, ed. D. A. Goslin (Chicago, Rand McNally, 1968), p. 961.

44. Milton Rokeach, "The Role of Values in Public Opinion Research," *Public Opinion Quarterly*, 32 (1968–69), 547–559.

45. Hubert Marshall, "Politics and Efficiency in Water Development" in *Water Research*, ed. Allen V. Kneese and Stephen C. Smith (Baltimore, Johns Hopkins Press, 1966), pp. 291–310.

**Thomas J. Nieman**

46. Everett Rogers, *Diffusion of Innovations* (New York, Free Press, 1962), pp. 311–313.

47. Joseph H. Schumpeter, *The Theory of Economic Development* (New York, Oxford University Press, 1961), p. 85.

TABLE 8.10

*Professional Practice Variables*

| Variables | Consulting Engineers (mean) | Public Health Officials (mean) | Difference (mean) |
|---|---|---|---|
| Age | 41.6 years | 55.2 | 13.6 |
| Professional experience | 15.6 | 20.4 | 4.8 |
| Years with present employer | 9.0 | 13.5 | 4.5 |
| Salary level | $22,500 | $20,000 | $2,500 |

This limited degree of mobility open to public health officials may contribute to their more conservative stance on the acceptance of renovated wastewater. Over time, public health departments tend to stabilize in size, and the individuals initially employed continue to retain their positions. Other qualified personnel are thus unable to gain access to positions of responsibility, and as a result public health officials are older and more experienced but with less mobility, status, and earning power than consulting engineers.

This has led to a situation where, *ceteris paribus*, conditions conducive to the adoption of innovations are more predominant in consulting engineering firms than in public health departments. Everett Rogers, in his classic *Diffusion of Innovations*, finds that early adopters are younger, enjoy a higher social status and more favorable financial position, and are more mobile than late adopters.[46] Although consulting engineers may not be entrepreneurs in the Schumpeterian sense (that is, risk-takers who are responsible for innovations),[47] at a minimum they appear to be in a better position to respond positively to the adoption of innovations.

In short, then, while both groups are of similar social origin, background, and educational level, they differ in their professional careers. Although it is not clear what motivates an individual to choose a particular profession, once he becomes part of that profession his attitude tends to resemble that which prevails in the profession as a whole. Thus in selecting a profession the neophyte is also selecting an attitude with which he can feel comfortable. It seems reasonable to conclude, therefore, that professional socialization plays some role, although the extent is unclear, in the formation of an individual's attitude toward the potable use of renovated wastewater. None of this should be taken to detract, however, from the sound professional concerns that lead public health officials or consulting engineers to question or oppose direct potable re-use of wastewater. This study simply suggests that there appear to be behavioral factors that reinforce or augment the professional concerns.

## Personal and Psychological Bases

Personal ideas and feelings not necessarily reinforced by the profession may also contribute to these attitudes. Because of the varying sociological

TABLE 8.11

*Attitude toward Personal Health*

| Attitude | Consulting Engineers | | Public Health Officials | |
|---|---|---|---|---|
| | No. | % | No. | % |
| Take health for granted | 50 | 51.0 | 6 | 27.3 |
| Pay attention to health | 48 | 49.0 | 16 | 72.7 |

$X^2 = 3.17$      Sig. at .10      N = 120

constitution of each profession, it is likely that each individual within the profession will hold certain values to be of greater importance than others within the value system. These values, in turn, relate directly to the attitudes developed toward specific objects and situations.[48] A value can be described as a standard by which to guide the formation of attitudes and which helps a person in choosing between alternatives to resolve situational conflicts. Since professions tend to attract individuals with similar attitudes, these personal ideas and feelings may partially explain the attitudes of the two groups. This research attempts to explain further attitudinal differences in terms of certain selected psychological dimensions: (1) fear of incorporation of, or contact with, the impure or dangerous, (2) progressive versus traditional stance, and (3) revulsion to body waste tendency. A sentence completion test then served to solicit responses along each of the psychological dimensions.[49]

The "fear of incorporation, or contact with the impure or dangerous" dimension relates to a situation in which an individual will express concerns over the incorporation of some substance that may be impure or dangerous. This concept is similar in nature to the idea of cleanliness as a value, in that an individual is concerned with coming into contact with unclean objects or substances. Previous research has demonstrated that some religious groups are more concerned than others with cleanliness.[50] The ingestion of renovated wastewater may be considered from the same perspective, as it is a substance, although purified, that has been associated with impure or dangerous (disease-causing) wastes. In relation to consulting engineers and public health officials, the rationale is that an individual expressing a fear of incorporating the impure or dangerous will be more likely to reject renovated wastewater as a viable water supply alternative.

To measure this dimension, the respondent was asked to indicate which of two statements ("I tend to take my health for granted and rarely give it a thought until something goes wrong and I don't feel well," or "I pay considerable attention to my health and am careful to take care of myself") more closely represents his attitude toward his personal health. The attitude toward one's health may be a constant situational problem in which care is taken to avoid, no matter how remote, any possible health hazard. On the other hand, personal health may be of little concern and the individual may be less cautious about coming in contact with the impure or dangerous. Whereas the consulting engineers divide about evenly between the two posi-

48. Rokeach, p. 551.

49. John H. Sims, Professor of Psychology at George Williams College, developed a sentence completion test to measure the seven psychological dimensions utilized in the study. These items have been pretested on 225 respondents in relation to public attitudes toward renovated wastewater and on ten consulting engineers prior to utilization in this study.

50. Milton Rokeach, "Faith, Hope, and Bigotry," *Psychology Today*, 3 (April 1970), 33–37, 58.

51. Rogers, p. 124.

52. Homer G. Barnett, *Innovation, the Basis of Cultural Change* (New York, McGraw Hill, 1953), p. 313.

TABLE 8.12

*Propensity to Innovate*

| Old Ways and/or New Ways | Consulting Engineers | | Public Health Officials | |
|---|---|---|---|---|
| | No. | % | No. | % |
| Prefer new ways | 42 | 42.9 | 5 | 22.7 |
| Prefer old ways | 26 | 26.5 | 11 | 50.0 |
| No particular preference | 30 | 30.6 | 6 | 27.3 |

$X^2 = 5.14$     Sig. at .10     N = 120

tions, nearly three out of four public health officials "pay close attention to their health" (Table 8.11). While this is not surprising given the nature of the profession, the personal concern with health may well reinforce or contribute to the prevalent negative attitudes in the profession.

The progressive-versus-traditional dimension relates to the propensity of an individual to innovate. According to Rogers, "innovativeness is related to a modern rather than a traditional orientation."[51] In this sense the reception given a new idea is not completely unpredictable, since the character of an idea is generally measured by its compatibility with prior cultural elements.[52] The idea of renovated wastewater for potable use, therefore, may be foreign to consulting engineers and public health officials who have traditionally utilized water from relatively unpolluted sources. In fact, public health officials are specifically instructed in water supply standards to use the highest quality water available. The rationale for utilizing this dimension, then, is to determine whether the willingness to accept renovated wastewater relates to a person's general willingness to innovate.

The results obtained from completion of the sentence stem "When old ways and new ways of doing something are in conflict, I usually find myself on the side of the . . ." measured this dimension. Twice as many public health officials as consulting engineers expressed a preference for the traditional "old ways" (Table 8.12). More of the consulting engineers (43 percent compared to 23 percent) demonstrated a more progressive outlook in choosing between conflicting methodologies by expressing a preference for "new ways." These differences in preference are consonant with the differing attitudes of the groups toward potable use of renovated wastewater. Although consulting engineers appear more progressive, they are not overwhelmingly so. These results do suggest, holding constant for a moment the issues of content, that consulting engineers are more likely to take the lead in proposing innovative uses of renovated wastewater.

The final psychological dimension, "revulsion to body waste," involves the tendency of being appalled at the thought of coming in contact with what may be considered body wastes. The psychological literature has not previously explored this dimension. Nonetheless, since wastewater consists partially of body wastes, this tendency may have some relevance to any system involving body contact with or ingestion of renovated wastewater. An individual with an aversion toward body waste would be likely to react negatively to-

TABLE 8.13

*Body Waste Revulsion Tendency*

| | Consulting Engineers | | Public Health Officials | |
|---|---|---|---|---|
| | No. | % | No. | % |
| *The Use of Untreated Human Excrement for Fertilizer*[a] | | | | |
| Acceptable | 27 | 27.6 | 4 | 18.2 |
| Neutral | 41 | 41.8 | 5 | 22.7 |
| Repulsed by the idea | 30 | 30.6 | 13 | 59.1 |
| *Astronauts' Sustaining Body Fluids by Drinking Their Own Urine After Purification*[b] | | | | |
| Acceptable | 73 | 74.5 | 10 | 45.5 |
| Neutral | 17 | 17.3 | 3 | 13.6 |
| Repulsed by the Idea | 8 | 8.2 | 9 | 40.9 |
| *Drinking Your Own Saliva Out of a Sterilized Test Tube*[c] | | | | |
| Acceptable | 40 | 40.8 | 7 | 31.8 |
| Neutral | 34 | 34.7 | 4 | 18.2 |
| Repulsed by the idea | 24 | 24.5 | 11 | 50.0 |

[a]$X^2$ = 6.38  Sig. at .05  N = 120
[b]$X^2$ = 15.94  Sig. at .001  N = 120
[c]$X^2$ = 5.93  Sig. at .10  N = 120

ward high-order uses of renovated wastewater, whereas an individual not particularly concerned about contact with body wastes would be more likely to respond favorably. The research instrument included three scaled items concerned with (1) using untreated human excrement for fertilizer, (2) astronauts' sustaining body fluids by drinking their own urine after purification, and (3) drinking one's own saliva out of a sterilized test tube.

The attitudinal difference revealed between the two groups may be the behavioral factor (but not including professional concerns) that accounts most for their differing attitudes toward renovated wastewater. Public health officials exhibit an apparent "revulsion to body waste" in that 59 percent are repulsed by the idea of using untreated human excrement for fertilizer, 40 percent by the idea of astronauts sustaining body fluids with purified urine, and 50 percent by the idea of reincorporating their own saliva (Table 8.13). The consulting engineers exhibit substantially less revulsion; respectively, their positions are 31 percent, 8 percent, and 25 percent revulsion on the three issues. The body-waste revulsion dimension, relating so directly to the potable use of renovated wastewater, calls attention to the possible rejection of water re-use systems, not on professional grounds but from aesthetic concerns (that is, the very idea that one is drinking what was once sewage is repulsive). This factor affects both groups, of course, but consulting en-

53. Duane D.
Baumann, *The Recrea-
tional Use of Domestic
Water Supply Reser-
voirs: Perception and
Choice*, University of
Chicago Department of
Geography Research
Papers, 121 (Chicago,
1969), p. 18.

54. James F. Johnson,
*Renovated Waste Wa-
ter*, University of
Chicago Department of
Geography Research
Papers, 135 (Chicago,
1971), pp. 50–52.

gineers appear more convinced that it is technologically possible to obtain
water safe for drinking from wastewater renovation outside the laboratory
situation. Public health officials as a group, on the other hand, express
greater doubt over this possibility and question the wisdom of the proponents.

Within the psychological dimensions, consulting engineers and public
health officials are similar in relation to their general trust in science and
technology—an outlook common to technologically oriented professions.
The differences that appear correlate with the more general conservative
nature of public health officials. These officials consistently reveal greater
conservatism on specific psychological constructs, a position that is consistent with their extremely conservative professional outlook toward direct
potable re-use. Consulting engineers tend to be less conservative, a propensity that contributes to or reinforces professional caution rather than
resistance to potable re-use.

## Regional Differences

This study suggests the existence of significant attitudinal differences toward
renovated wastewater between consulting engineers and public health officials as professional groups. It is also possible that attitudes may vary geo-
graphically. Does the fact that professionals were reared, educated, and
presently live and work in a particular geographical region have any possible
significance for attitudes toward the adoption of renovated wastewater?

There is evidence to support the notion that regional attitudes may indeed
exist. In a study on the recreational use of domestic water-supply reservoirs,
Baumann noted differing regional attitudes.[53] For example, officials prohibit
almost entirely recreation on water-supply reservoirs in the Northeast and
Far West, but elsewhere in the United States, permit it. Within the Northeast,
the New England states have the most restrictive policies. In the Far West,
Washington and Oregon are the most restrictive. The professionals and
water managers in these regions, more so than in others, do not perceive
water-oriented recreation on domestic water-supply reservoirs as being
compatible with safe, potable water supplies.

In relation to renovated wastewater, James Johnson speculates that potable re-use would be unacceptable in the Northwest and Northeast and acceptable in the Southwest, Midwest, and Great Plains.[54] He justifies this possible regionalization of attitude on the premise that inhabitants of arid
environments, especially if water shortage is experienced, would be most
likely to accept renovated wastewater. In addition, Johnson found that the acceptance of renovated wastewater correlated with the degree of pollution
perceived in the existing source of supply. It may be that, in regions where
the water supply originates from a polluted source, the use of renovated
wastewater as an alternative resource is more likely to be accepted than in
regions where the source of the water supply is relatively unpolluted.

On the basis of previous studies, the sample of consulting engineers and
public health officials recognized three geographical regions; the East, the

TABLE 8.14

*Overtly Expressed Attitude toward Direct Potable Re-Use: By Region*

| Personal Feelings Expressed | East | | Midwest | | West | |
|---|---|---|---|---|---|---|
| | No. | % | No. | % | No. | % |
| For recycling: no qualifications | 3 | 7.5 | 15 | 35.7 | 3 | 7.9 |
| Hesitant to accept: economics | 4 | 10.0 | 13 | 31.0 | 6 | 15.8 |
| Hesitant to accept: operations | 6 | 15.0 | 6 | 14.3 | 17 | 44.7 |
| Hesitant to accept: technology | 18 | 45.0 | 5 | 11.9 | 7 | 18.4 |
| Hesitant to accept or reject: aesthetics | 9 | 22.5 | 3 | 7.1 | 5 | 13.2 |

$X^2 = 40.41$      Sig. at .001      N = 120

Midwest, and the West (Table 8.1). Since direct potable re-use is not utilized in the United States at present, the study chose regions that were most likely to manifest the greatest differences in attitudes toward using renovated wastewater. The anticipation was that attitudes would be most negative in the East, less negative in the Midwest, and least negative in the West.

To elicit an overtly expressed attitude representative of the three selected regions, the interviewer asked respondents to state directly their personal feelings as to the possible implementation of renovated wastewater for potable use. The opinions expressed (Table 8.14) indicate that most respondents hesitate to endorse the idea, but the reasons vary by region. In the East a significant percentage (45 percent) of the respondents hesitate on technological grounds, while another 28 percent are concerned with aesthetic factors. Midwestern respondents tend to accept without qualifications (36 percent) or to hesitate because of economic feasibility (31 percent). Respondents in the West (45 percent) express concern chiefly for the adequacy of plant operations (that is, the possibility of human error affecting the re-use process).

The attitude of the respondents was also projected onto others in the form of personal suggestions regarding the resolution of a water shortage by use of renovated wastewater. The most positive attitude appears in the Midwest, where 29 percent responded affirmatively to the use of renovated wastewater to solve the water shortage (Table 8.15). The Eastern respondents, with 43 percent accepting only as a last resort, are the most negative. Uncertainty rather than a very positive or negative attitude seems to characterize those in the West. This projection of attitudes is consistent with the overt attitudinal expression shown in the previous table.

Generally, the regional variation in attitude can be summarized as follows: in the East attitudes appear to range from very conservative to slightly conservative, with technological and/or aesthetic considerations the major concerns. The Midwestern attitude is more liberal in the sense that renovated wastewater appears to be much more acceptable and is a likely possibility if proven economically feasible. In the West the predominant attitude is one of caution in regard to the manner of its use.

TABLE 8.15

*Projected Attitude toward Direct Potable Re-Use: By Region*

| Projected Attitude | East | | Midwest | | West | |
|---|---|---|---|---|---|---|
| | No. | % | No. | % | No. | % |
| For recycling: no qualifications | 5 | 12.5 | 12 | 28.6 | 7 | 18.4 |
| For partial use and other uses | 3 | 7.5 | 4 | 9.5 | 11 | 28.9 |
| Uncertain and postpone decision | 9 | 22.5 | 7 | 16.7 | 10 | 26.3 |
| Use only as a last resort | 17 | 42.5 | 8 | 19.0 | 2 | 5.3 |
| Reject: no qualifications | 6 | 15.0 | 11 | 26.2 | 8 | 21.1 |

$X^2 = 24.44$     Sig. at .01     $N = 120$

## Summary and Conclusions

This study has identified consulting engineers and public health officials as professionals whose influences play a major role in municipal water-resource planning and whose attitudes may profoundly affect the future of water re-use for American cities. Although attitudes toward the potable use of renovated wastewater vary, neither group is willing to commit itself at this time to the adoption of the process.

Public health officials have openly expressed and projected a definite reluctance to accept renovated wastewater for potable use. This reluctance stems, of course, in large part from the professional concerns over health issues discussed at length in Chapter 4. But this study suggests that there may be behavioral reasons as well. In sociological terms, for example, this negative attitude appears to relate to the fact that public health officials are older, more experienced, and less mobile individuals with somewhat lower earnings than consulting engineers. These characteristics probably enhance a tendency to maintain the status quo. This apparent conservatism also invades the psychological dimensions, in that public health officials are traditionally oriented, fear coming in contact with or incorporating what may be potentially dangerous, and express, to a large degree, revulsion to body wastes.

The willingness of consulting engineers to accept the concept of direct potable re-use but their hesitance to adopt it at present correlates with the fact that they are younger, somewhat less experienced, and more mobile individuals with greater earning power than public health officials. These are all factors conducive to innovation. Within the psychological sphere, there clearly is conservatism among the engineers, who tend to be slightly traditional, somewhat concerned about coming into contact with, or incorporating, what may potentially be impure. This chapter also suggests a more favorable professional climate for re-use innovation in the West and Midwest. A strong resistance is characteristic of the East, a generally positive but mixed attitude is apparent in the Midwest, and an attitude of cautious

TABLE 8.16

*Interaction of Attitudes: By Region and Profession*

| | | |
|---|---|---|
| Public Health Officials ⟷ Consulting Engineers | | |
| Negative Attitude | | Moderate Attitude |
| East ⟷ Midwest ⟷ West | | |

acceptance characterizes the West. Table 8.16 summarizes the interplay between profession and region.

Unlike the findings of the two chapters on economic feasibility, the results of the research in this chapter do not lead to optimism for early progress with high-order, and especially potable uses, of reclaimed wastewater. In this sense it is consistent with the findings of Chapters 4 and 7. At least, re-use innovation must first overcome the very cautious stance of the consulting engineers and the outright resistance of public health officials. Both groups stand to lose a good deal, of course, in assuming strong positive leadership in the innovative process. Yet these attitudes may be conducive to change as technology develops, as safety issues are resolved, and as experience with water re-use accumulates. In any event, it is necessary to assess the entrance of these attitudes into the decision process. Only then will the impact of attitudes upon the future potable use of renovated water in American cities become clear. This is the task of the next chapter.

# Part III. Water Re-Use, Innovation, and Public Policy

# The Water Management System and the Reclamation Innovation
David McCauley

**9**

Part II of this book analyzed at length the issues that may potentially obstruct the diffusion of water re-use systems in urban America over the next several decades. The preceding chapter centered particularly on the reservations of managers and the bases of their attitudes. It is appropriate now to analyze the interaction of a number of these elements in the process of innovation as a whole.

In this chapter the foci are the water manager's images of the acceptability of reclamation and the managerial experiences with specific wastewater reclamation projects. These issues are presented within a framework model of the water-management system. Many ideas raised elsewhere in this book are connected to, and clarify, this general model.* The model describes a process that is repeated by different sets of water managers, and to this extent the chapter is iterative. On the other hand, it does not attempt to recapitulate details of site descriptions which may be found in Chapter 3, above.

## Introduction to the Model

*The adaptive system*. Political and administrative units are adaptive systems. They are born, energized, or transformed as they adapt to problems in their environments. But these systems do not react passively; each organization has its own goals and *strategies*.[1] The interactions of these organizations, with one another and with their relevant environments, generate patterns of tremendous complexity. Comprehensive water-resources management is just one example.

The goal here is to analyze the decision process. This analysis entails at least three aspects:

(1) to identify certain elements of the decision process:
   (a) those which are primarily administrative or political
   (b) the degree of intention and planning
   (c) the intercomponent coordination
   (d) the impact of environmental need (crisis response, limited range of choice, pollution vs. water supply)
(2) to identify the important variables and resulting strategies of individual actor types (for example, politicians and water managers)

*The model that follows is frankly exploratory. The number of water managers interviewed is small (approximately thirty-five). In few of the sites was there major political activity available for study. In other possible sites it was of dubious value. For example, in Chanute, Kansas, the conditions for implementing re-use were very difficult; as a study site it is therefore questionable. In exploring some of the re-use experiences, the basic method was to ask managers to state what they would do in situations whose very real political pressures they could not feel. The reader should keep these points in mind as he reviews the study.

1. "Strategies" as used here means "a combination of means, ends, and decision criteria." See Gilbert F. White, *Strategies of American Water Management* (Ann Arbor, University of Michigan Press, 1969), p. 9.

(3)  to analyze each site through the decision model, in order to determine the similarities and differences.

Many studies of organizations or decision processes treat these elements somewhat in isolation. The present study will deal with specific organizations interacting with specific environments. The water-management system can be seen as an adaptive system encountering a novel situation. It is innovating in the sense that it attempts to use a new strategy in

(1)  dealing with old problems (water supply acquisition or pollution control)
(2)  dealing with a new or compounded problem (how to deal with both water supply and pollution control).

The system is innovating because the traditional responses to environmental problems are no longer suitable. The inapplicability of traditional strategies may result from environmental stringency (constriction of the usual range of environmental opportunities) or from a change (emphasis on efficiency, new personnel) in the system attitudes or roles. In any event, the system is being studied as it undergoes some strategy shift.

This directs attention to certain areas:

(1)  the quantity and variety of information in the system
(2)  the structure of the communications and decisions networks
(3)  the pattern of the subsystems within the whole
(4)  the operating rules guiding the system's structure and behavior.

*Definition of the water-management system*. The water-management system is a man-technology unit: it is the people and technologies significantly involved in the provision of input, movement of throughput, and modification of output (with water as the commodity) through a given system (in this case a municipal system). Given an interest here in policy formulation, the focus is on the people. The technology can be understood as a set of characteristics (efficiency, complexity) which must be manipulated and as a set of problems (hazard potential, integration into system design) which must be solved.

Included in the strict definition of the water management system are those having either formal authority or direct influence on water resources decisions: water supply personnel, consulting engineers, water commissioners, and the chief executive and city council. This group then interacts with a social or legislative environment which includes citizens' groups, more specific interest groups (for example, industry, agriculture), and relevant state or regional agencies. The technological interaction of this group with the biophysical environment is implied.

Furthermore, the water-management system is seen as one subsystem among many vying for the attention of municipal officials. Its position of one among the many has important consequences for the type of planning and amount of attention it receives.

The present authors are seeking to determine how and why the water-management system innovates, what problems it encounters, and what strategies it adopts to deal with these problems.

*Water-management system—a three-state description*. It is helpful to view
the reclamation innovation as a three-state process (Table 9.1). Different
components (technical, political, community) unequally perceive the importance and inclusion of environmental variables, although some may be very
important to all. At different points in the innovation process a given component is more active than at another point and has different priorities.

Each state includes a certain set of actors, problems, and strategies. But it
is important to stress that these states are not independent of one another.
Some of the actors, problems, and strategies exist in two or all of the states,
and relatively constant suprasystem elements constrain each state. Yet in
the movement from one state to another some problems drop out or become
less important, some patterns of interaction change, and some strategies
change. Moreover, different components of the system assume responsibility for different aspects of a decision, and these concerns endure through the
decision process. Different actors also define their roles in their own ways
(for example, "professional" concerns), although the extent of difference varies widely.[2]

Table 9.1 shows a three-state system operating under a certain number of
constraints. Obviously, if the actors, problems, and strategies vary somewhat with each state, so too will the constraints. Figure 9.1 is a more detailed
diagrammatic presentation. The process is not as discrete, however, as the
formal model would imply. Each state deals with a certain thrust or constella-
tion of activities, but does not represent a strict disjunction. Despite its three
separate states, the process is in many dimensions continuous.

Indeed, the same actors or problems may be present in two or all of the
states. Overtly, utility department personnel may try to affect public opinion
at all stages of the process; challenges to legal and political constraints may
occur throughout the whole process. Some of the continuous influences are
more subtle. Formal interaction with the Department of Public Health occurs
during State II. Expectations regarding Public Health Department behavior,
however, may affect the decision process from the start. The same is true of
perception of public attitudes. Even though formal interaction occurs in State
III, the manager's perception of public attitudes might elicit initial strong support of re-use or, alternatively, prevent even the proposal of the idea.

2. This definition of
concern is obviously not
absolute. It depends, for
example, on how tightly
people define their
roles. It may also de-
pend on the nature of
the concern (for exam-
ple, nearly all actors feel
able to comment on
political or public-
attitude aspects of a
problem; fewer will
comment on economic
feasibility, and fewer
still on technical feasibility).

3. See Roger Kasper-
son, "Political Behavior
and the Decision-
Making Process in the
Allocation of Water Re-
sources between Rec-
reational and Municipal
Uses," *Natural Re-
sources Journal*, 9
(1969), 188.

## The Technical Component (I)

*Role of the technical component*. The water-management system as a
whole is aware of the general water supply and water quality situation. Although there are widely differing levels of information and interest, the various system components are usually cognizant of the immediate alternatives
(for example, a certain well field is being investigated or a certain river is
being studied). These images, however, are often very fuzzy and in some
cases are without foundation.[3] A technical unit (department personnel
and/or consulting engineers) carries on the more specific evaluations and
investigations. Such resource analysis may entail close interaction with in-

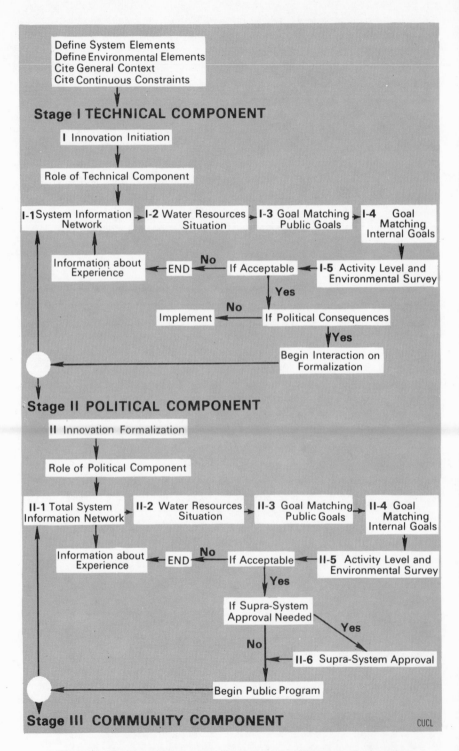

Figure 9.1: The Water Management System and the Wastewater Reclamation
Innovation: A Flow Model.

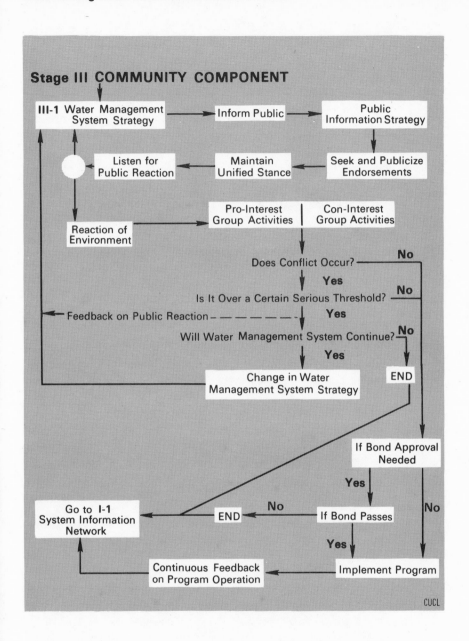

**Stage III COMMUNITY COMPONENT**

III-1 Water Management System Strategy → Inform Public → Public Information Strategy

Listen for Public Reaction ← Maintain Unified Stance ← Seek and Publicize Endorsements

Reaction of Environment → Pro-Interest Group Activities | Con-Interest Group Activities

Does Conflict Occur? — No

Yes

Is It Over a Certain Serious Threshold? — No

Feedback on Public Reaction — — — — — — Yes

Will Water Management System Continue? — No

Yes

Change in Water Management System Strategy     END

If Bond Approval Needed

Yes

Go to I-1 System Information Network ← No ← END ← If Bond Passes     No

Yes

Continuous Feedback on Program Operation ← Implement Program

CUCL

TABLE 9.1

*Organization-Environment-Interaction: A Three-State System*

*Technical Component (Innovation Initiation)*
   *Actors*: department personnel; consulting engineers.

   *Major Problems*: preliminary resource situation analysis; role of system
   bias (institutional, historical, public attitude perception); source and
   meaning of information inputs; preliminary economic and technical
   analysis.

   *Strategies*: convince political actors; convince public groups; assess
   impact on system design; assess use level of reclamation.

*Political Component (Innovation Formalization)*
   *Actors*: department personnel (with consulting engineers) interact with
   appointed and elected officials; Department of Public Health and other
   suprasystem representatives; some interest groups.

   *Major Problems*: costs and benefits from different viewpoints; variations
   in environmental need and perception; competing subsystem demands;
   problems in economic-technological analysis.

   *Strategies*: evolution of united stance; public opinion programs and role
   of media; creation of political system strategy.

*Community Component (System-Environment Interaction and
Consolidation)*
   *Actors*: system representatives; pro-groups, con-groups (fringe individ-
   uals); uncommitted public, system clients.

   *Major Problems*: community conflict potential; evolution of conflict; role of
   public attitudes; ecology of games.

   *Strategies*: assess progress of conflict and form policy; attitude change
   programs; level of reuse and associated conflict potential.


terested nontechnical people or may underscore a strong water policy commitment (for example, persistent support for water-supply increments) with
the technical component assigned "the job of how to do it."

   This assessment role of the technical component is of paramount concern
in the context of innovation in the reclamation of wastewater. It is almost
invariably this technical group which initiates the reclamation decision. In the
study sites analyzed herein, these technical people (two wastewater professionals, one water superintendent, and a city planning group) both proposed
and championed the idea.

   Even if the idea were to enter the water management system through
another channel, the technical role would be important. Technical evaluation
of the feasibility of a proposal would greatly influence, often determine, its
prospects for adoption. This is particularly true if the professionals' opinions

are respected (as they were in the study sites) by other members of the water-management system.

Efficient communication is an essential element in the innovation process. This communication system must transmit without undue distortion many types of information; to function effectively, the system must be able to carry a wide variety of data. Attention must be directed to the terminals and nodes of the system as well as to the adequacy of the links; for the nodes are switching and transforming devices and as such can greatly affect the information carried. The technical element is one such node. An analysis of the wastewater reclamation decision can properly begin with the problems and behavior of the technical component.

*System information network (I-1)*. How did the idea of wastewater re-use as a management technique enter into and gain initial support in the water-management system? Here one must distinguish between the "idea" and the "plan" (or the necessary implications of the idea). The transition from idea to plan comes with clarification of objectives, determination and evaluation of means to the objective, and development strategies.

The idea of reclaiming wastewater for further uses was not new to the study sites in question. As Chapters 2 and 3 make clear, Texas sites had used municipal effluents for agricultural purposes for many years (over thirty years in Lubbock and almost fifty in San Angelo). This symbiotic relationship had been arranged long ago in contracts with local farmers who wanted the water. The objective of the water-management system, however, was primarily to use ground-spreading as a means of effluent treatment rather than as an innovative effort toward water re-use. Such examples of inadvertent re-use are qualitatively different from the on-going re-use efforts in our study sites, where reclamation and re-use per se are the objective (see the distinctions in Chapter 1). The same was true for industrial uses. Respondents in Texas sites could cite numerous examples of this use. Many commented on the frequency of the practice in the Southwest and its infrequency in the East. They saw these uses as a function of environmental necessity involving only minor technical problems.

For higher-order uses where problems are more complex, information levels were also high.[4] Lubbock, Tucson, and Antelope Valley officials all had extensive information on Santee, California, which was their important model. When it became clear that re-use was a good possibility in their communities, they visited Santee. Santee, during its early experimentation, had knowledge of the ornamental lakes program at Golden Gate Park and of the total reclamation system at Camp Pendleton Marine Base. The Santee managers appeared to have the most extensive information on reclamation of all the sites. In San Angelo, where reclamation for drinking was a possibility, the managers were aware of the Chanute, Kansas, and Windhoek, Namibia, experiences in planned, direct potable re-use. Although managers in other sites knew of these examples, their levels of information, with the exception of Santee, appeared to be lower than those in San Angelo.

The idea of re-use is a familiar one in the water-short environment of the arid states, and all of the managers interviewed endorsed it for lower-order

4. Since the study investigated sites where re-use was under consideration, one would naturally expect awareness of other re-use experiences to be high. It may also be high in those sites where, although water re-use is not being considered, highly qualified personnel are present. The awareness among other components of these latter systems, however, would not be as high.

5. See, for example, Robert Fleming and Harold Jobes, "Water Reuse: A Texas Necessity," *Journal of the Water Pollution Control Federation*, 41, (1969), 1564–69.

6. See editorial and news story, *Antelope Valley Press* (August 30, 1962).

7. Similar information networks suited to their own needs exist for other components of the system (for example, political). The professional network, however, is probably better organized and carries more reliable information.

uses and agreed that it was inevitable.[5] There was considerably less agreement, as the preceding chapters make clear, on higher-order uses.

The main point in initiation of innovation, however, is not merely awareness of the concept but, rather, adequate levels of specific information on the idea and a fairly clear image of how the idea relates to the environment of the organization. The important person or group is not the one who merely proposes the idea, but the unit that actively supports or champions it (although these are frequently the same people) and which satisfactorily identifies certain applications.

In the study sites, the technical component conducted this initial clarification. It was greatly aided by its place in an information network with both intra- and inter-system links. As water-management professionals, these people found that outside links were particularly important in providing technical information for the re-use decision.

This discussion stresses the initiatory role of the technical component. The Antelope Valley Project, an ornamental, low-contact recreation lakes scheme, is an interesting exception. Here the process appears to have been initiated by a political actor who urged the adoption of a Santee-like scheme.[6] He could do this because the initially large technical problems had been solved at Santee and had become correspondingly less important. Consequently, he could endorse the transplant of the Santee Model. As technical ambiguity decreases, one can expect the other components to play a more important role in the early stages.

To water-management professionals, the outside or intersystem links are very important in the reclamation and re-use decision. More information about a given alternative can be obtained from this network. One can to a great extent identify the attributes: costs, technical requirements and potential, problems and hazards, public reaction, lessons from experience—in brief, the costs and benefits of an alternative.[7]

This network comprises both formal and informal elements and a wide variety of sources: published literature (journals, books, technical reports, consultants' reports, etc.), informal communications (letters, discussions), formal communications (professional associations, meetings), and informal person interaction. Although the present research undertook no rigorous examination of the composition of the linkages of the managers in the study sites, it was obvious from informal observation and their high information levels that their links were extralocal and adequate. The size and complexity of the information network also tended to expand as the innovation progressed. At Santee the links went from effectively regional to national and international. In San Angelo the development of linkages to federal levels contracted as the reclamation potential diminished.

The stances of professional associations and colleagues are important referents for water managers. Managers supporting wastewater reclamation reported that their colleagues tended to regard reclamation positively. Chapter 8 suggests that even apparent agreement between one's activities and these group opinions is very helpful. These norms may also yield personal satisfaction in the sense that one may feel innovative. The information on norms, techniques, and other elements (for example, public reaction) can have the effect of reducing the novelty associated with an alternative and increasing the level of uncertainty. It tends to make the image of an alternative seem less chancy, less experimental, and this can reduce the personal reservations of the water manager and those of the actors in the other components.

The local, or intrasystem, information network provides information on the characteristics of the immediate environment. Such factors as the expected public reaction, the posture of the political components, the position of public health officials, and the actual and potential demands of clients are important elements in the re-use decision, and adequate information on them is essential. Even though the central task at this stage in the innovation process is technical, it would be a mistake to evaluate the situation solely from a technical perspective.

The local information was generally obtained informally. Methods included individual and small-group discussions, feedback from talks to community organizations, "coffee meetings," media, and evaluations of system performance from both clients and the general public. The strategies can differ. In San Angelo and Colorado Springs, feedback at the community organization meetings was important. In Santee, "coffee klatsches" and "experiments," such as developing the lakes to a given level and then waiting for public response, were effective techniques.

By the time that formalization of the interaction with the political component begins, the technical component has amassed a good deal of information on a wide range of system elements. An important part of this process is the identification of potential allies in other components and the strengthening of links with them. In one situation (San Angelo), these links might be to the mayor; in another (Lubbock), certain councilmen might be important. The Chamber of Commerce might also play a linkage role.

Here, then, is a water-management system whose technical component possesses a certain level of information on wastewater reclamation. What water resource situations stimulated the organization to seek this information? What environmental conditions in the water-management system encourage consideration of wastewater re-use as a realistic management alternative?

*Water resource situation (I-2).* The improvements in effluent quality in the study sites resulted from the need to meet standards imposed by political agencies. Only one site increased effluent quality specifically for re-use. Santee, in a prolonged program of experimentation for scientific (for example, virus studies) and recreation purposes, directed its efforts toward reclamation.[8] In other sites (for example, Colorado Springs), pilot projects in advanced waste treatment for re-use were under way, but increments in their larger facilities were not used for this purpose.

Elsewhere, the water-quality standard induced an increment in effluent which offered the opportunities for reclamation (rather than reclamation opportunities being the stimulus for increases in effluent quality). In some

9. See the discussion in Ernest Flack, "Urban Water: Multiple Use Concepts," *Journal of the American Water Works Association*, 63 (1971) 644–646.

cases the reclamation uses required only low-quality water; this was the case in the early irrigation experiences at Colorado Springs. At Lake Tahoe, Alpine County standards were responsible for the quality increase, and effluent was for some time underutilized in terms of quality (it could have supported swimming). The City Commission in San Angelo wanted studies on four levels of quality improvement, one of which would provide for reclamation. In summary, possible re-use benefits were identified (sometimes fairly early) as standard-induced increments were discussed, but they were not the reasons for the improvements. Re-use was a stimulus, but an insufficient one.

The water-supply possibilities were, of course, valued much more than were the water-quality improvements *per se*. The water-supply benefits stayed in the community, whereas the water-quality benefits were perceived to accrue to others. Reclamation served three distinct but not mutually exclusive types of water-supply purposes. The first was an expansion of traditional water-supply services. Industrial uses of effluent in Lubbock and Colorado Springs are examples. A second application was for special-purpose functions. The ornamental and recreational lakes in Lubbock and Santee exemplify this type. A third, less distinct, type allows uses that would otherwise be excluded or put at a cost disadvantage. Municipal golf-course and park irrigation are in this category.

Reclaimed wastewater for lakes is a very prominent application. Citation of the Santee example in proposed lake projects in Lubbock, Antelope Valley, Tucson, and East Lansing suggests that its utility is widely recognized. The advantages of re-using wastewater for industrial expansion is also widely recognized and highly valued. The provision of low-cost water for marginal activities is also widely appreciated but less highly valued.

These applications highlight the relationship of re-use to need. Ornamental lakes, for example, are pleasant but not essential. Indeed, in all but one site (and in this site it was not adopted), the relationship of re-use to environmental need was very tenuous. Only two sites considered re-use in the context of immediate need. San Angelo, because of supply failure, considered and rejected planned, direct-potable re-use. Colorado Springs, during a dry period, installed hookups for irrigation. The difference in intended level of use explains much of the variation in these two sites.

But the re-use adoptions in Colorado Springs and Lubbock have not affected the movement toward source expansion.[9] The tendency to exploit new sources appears to be endemic. In the study sites, then, the question of need as a factor in re-use adoptions is ambiguous. The site where the immediate need was greatest failed to adopt re-use. The sites where projected needs were the rationale adopted re-use, but this had little effect on tendencies toward expansion of sources. Indeed, reclaimed water was applied in some of these cases for nonessential water-supply purposes (lakes). But nonessential use is different from non-valued use. Perhaps it is more appropriate to use the framework of client need rather than the needs of the system as a whole. The role of special clients' needs is outlined below in this chapter. Its impact upon the re-use adoption is considerable, particularly in the case of lakes and industrial applications. The apparently diffuse benefits of an ornamental lake can, upon consideration, be more accurately localized.

The role of crisis in water management is important.[10] In the sites under consideration it was ambiguous. San Angelo, in a crisis, rejected re-use; other sites, under less pressure, adopted it. Yet, in a sense, two sites (San Angelo and Colorado Springs) both have encountered the failure of traditional means of supply augmentation. One was immediate; one was projected. The water managers in San Angelo had few immediate options. The water managers in Colorado Springs saw that their present strategy (of intermontane diversions) would be blocked in the future. It renders the notion of crisis trivial to say that the restriction on future options of Colorado Springs constitutes a crisis, but it does make the useful point that in most of the study sites water reclamation and re-use were understood in the context of the present or projected failure of traditional strategies.

Early experiences with re-use tended to be accidental; later experiences, in the *same* and *different* sites, tended to be planned. Colorado Springs is a good example. Having started its nonpotable system with accidental hookups, it now plans replacement uses to release fresh water.

Each of the study sites represents a type of implementation. In Colorado Springs, the re-use innovation provides ground irrigation and industrial water supply. The reclamation program, started during a temporary period of stress, was handled informally by the water manager and certain clients. It now involves the planned satisfaction of need for a large volume of low quality water. Now part of a long-range, planned program, this mode of re-use is an extremely rational option, an example of the type of re-use system advocated throughout this book.

In Lubbock the Canyon Lakes Project is basically an amenity use of water. Reclamation serves a source of industrial water supply, but the lakes constitute a more prominent use. In the hands of the Planning Department, reclamation is a tool in a comprehensive, multipurpose planning scheme. Re-use here is not directly related to environmental need but, rather, meets the demand of certain clients for an amenity resource.

A similar situation prevailed until recently at Santee. Although the relationship to environmental need is hard to establish, re-use served as an amenity resource, although some water for irrigation was provided. Client satisfaction was an important stimulus here, but the innovation had its own dynamics as well. The policy of incremental levels of use evolved in a context where experimentation for scientific and recreational purposes was a major objective.

San Angelo had a crisis situation, and the intended use was potable. In a number of respects the situation resembled that of Chanute, Kansas. San Angelo did not adopt reclamation for re-use primarily because the intended use level (drinking) was too high—that purpose would have created too many problems. But had environmental stringency not abated, the outcome might have been different.

The relationship of reclamation to a more immediate definition of environmental need is ambiguous. It becomes somewhat clearer if one tries to relate it to client demand. The willingness to satisfy these demands is related to certain public and private goals which water managers support.

10. Henry Hart, "Crisis, Community, and Consent in Water Politics," *Law and Contemporary Problems*, 22 (1957), 510–537. See also Roger E. Kasperson, "Environmental Stress and the Municipal Political System" in *The Structure of Political Geography*, ed. Roger E. Kasperson and Julian V. Minghi (Chicago, Aldine, 1969).

11. Flack "Urban Water," describes it as "wanting to supply safe and adequate water to his customers at all times."

12. Gore's term. It is the "organizational goals describing the organization's relationship to the environment and what it expects to do for its environment." See William Gore, *Administrative Decision Making* (New York, Wiley, 1964), p. 76.

13. See the discussion in Jack Hirshleifer, James C. DeHaven, and Jerome W. Milliman, *Water Supply: Economics, Technology and Policy* (rev. ed., Chicago, University of Chicago Press, 1969).

14. Some implications of the commitment to avoid shortages are identified in C. Russell, D. Arey, and R. Kates, *Drought and Water Supply* (Baltimore, Johns Hopkins Press, 1970).

15. Baumann studies recreation and water supply in this context. See Duane Baumann, *Recreational Use of Water Supply Reservoirs*, University of Chicago Department of Geography Research Papers 121 (Chicago, 1969).

16. See the treatment of the benefits of wastewater reclamation in Paul Bonderson, "Quality Aspects of Wastewater Reclamation," *Journal of Sanitary Engineering*. ASCE, SA5, 90 (October 1964), 1–8.

*Public management goals (I-3)*. The general public goal of a water-management system is to deliver a high-quality water in sufficient quantity at a low price.[11] This is the agency's "mission conception."[12] But this goal is general; the prescriptions and proscriptions flowing from it are not severe. When searching for new water supply, one generally avoids polluted sources, or sources perceived to be polluted, but the finished product is clearly not as clean as possible. Also, managers do not appear to strive for maximum economic efficiency in water management (for example, a changed price schedule can delay the next increment).[13] Managers do, however, actively strive to meet their quantity goals. Restrictions are last resorts and shortages are anathema, even though they appear to be possible management tools.[14]

It is possible, then, to identify at least three separate goals: quality, price acceptability, and quantity. Although managers seek to achieve a certain acceptable level for each one, they tend not to be maximizers. In defining managerial response to wastewater reclamation, one must assess whether managers have perceived it as consistent with these goals. As Baumann has suggested, this goal-matching process is not a simple, completely rational exercise;[15] nor does it involve only public goals, for internal system goals must also be met.

Of central importance is the perceived relationship between wastewater reclamation and water-supply augmentation. Virtually all water managers interviewed emphasized quantity aspects more frequently than, for example, water-quality improvement benefits.[16] This is, of course, not surprising; it is an expression of the system's self-interest. Intercommunity altruism on water-resource questions has seldom been high.

Many of the managers recognized the potential management flexibility of wastewater reclamation. Briefly, they saw re-use as facilitating one or more of the following strategies:

(1) total supply augmentation—either as an addition to the municipal supply for unrestricted use (as in San Angelo, Texas) or for specialized lower-order use, thus releasing potable supply (as in Colorado Springs and other sites),

(2) a special purpose supply—for example, recreational lakes with specified activities as in Santee and Lubbock, or

(3) provision of supply for water-dependent activities which would presumably be otherwise excluded (for example, golf-course irrigation, park irrigation).

Wastewater re-use advances the public management goal of water-supply sufficiency. Whether the supply adequacy was seen from the point of view of the whole system or primarily that of a special client (for example, the Southwestern Power Plant in Lubbock, Texas), the water managers saw this goal as being served. This would be particularly true if there were few other management options. They perceived less broadly the potential of wastewater re-use as a drought-relief measure.

Environmental demands on the water-management system heightened the perception of the utility of wastewater reclamation for water-supply augmentation. The environmental demand was usually a projected one. In the study sites, supply adequacy is projected until approximately the year 2000. The "need," therefore, is a future need; yet in each of the sites the reclaimed effluent was described as essential and needed. In Lubbock there was competition for the effluent. Reclamation did not visibly affect the tendency to exploit new sources, and "water imperialism" continued. The reallocation (even apparent expansion with reclamation) of existing supplies may for some time be at a disadvantage relative to source expansion.

Wastewater reclamation was also seen as consistent with the low-cost goal. Some cost figures on reclamation were available, but they are complicated by ambiguities surrounding distribution, direct piping, dual systems, etc.[17] The important point is that managers regarded reclamation as being economical. As Chapters 1 and 5 make clear, the relative advantage of re-use markedly improves in two contexts. First, the cost of reclaimed water is regarded as the incremental cost, after the waste treatment costs to meet effluent standards are subtracted. This seems a logical policy because, managers argue, these costs would have to be incurred in any event. Secondly, the costs of reclamation are compared to the costs of increasingly inaccessible alternatives. The reduction of options acts, then, to increase the relative advantage of re-use. Furthermore, supramunicipal and regional agencies pay shares, some over 70 percent, of the capital costs of waste facilities. In the reclamation context, this amounts to a water-supply subsidy.

Related to both cost and supply aspects was the perceived avoidance of "waste" in re-use. Managers, particularly in areas of water need, expressed disapproval of using water once and then "throwing it away." At least one manager remarked on this as particularly bothersome when it involved water obtained through long-distance interbasin transfers, but many managers mentioned this aspect of waste in general. Why is not clear. Many managers remarked on the high value of water in their areas, at the same time that water was being underpriced. Others regarded this "waste" as inefficient, even as they were drawing water from projects of questionable efficiency. What appears to be involved in the avoidance of "waste" is the pursuit of an image of efficiency in an economic and technological sense, rather than as one example of a comprehensive attempt to manage water efficiently. This image may be due to an evolving environmental frame of mind, or it may reflect a particular orientation toward water. In any event the appearance of efficiency is a desirable public image to project. In Colorado Springs this image even enjoyed legal sanction. One compact regarding transmontane diversions specified that the water be re-used.

The question of the perceived consistency of wastewater reclamation with the quality goal is more complicated, but for unrestricted use (specifically for drinking) the matching is generally negative. Promoting reclaimed water at public demonstrations and endorsing it for use in municipal supply are two different situations, although one water superintendent did both. Though some water managers argued that their system's effluent was good enough to drink, they were reluctant to implement it as a municipal supply. The one

17. James Johnson does a cost comparison for wastewater reclamation at three sites: Tucson, Indianapolis, and Philadelphia. See Johnson, *Renovated Waste Water*, University of Chicago Department of Geography Research Papers, 135 (Chicago, 1971).

18. To what extent this assessment would vary with the method of inclusion in the system (for example, ground-water recharge, river mixing reservoir dilution) was not determined. The methods of inclusion are frequently limited locally anyway, and a preferred method (for example, ground-water recharge) may not be possible or practical.

19. Merrell, et al.

case in which planned potable re-use (through river mixing just above the municipal intake) received support involved a crisis situation, although not so serious as that of Chanute, Kansas.

Though the managers expressed confidence in the present and expected future developments in advanced wastewater treatment technology, they were less convinced of its day-to-day, in-service effectiveness. Operational questions such as continuous operation, input variations, and others were deterrents to direct potable re-use from a more strictly technical point of view. The public health questions discussed in Chapter 4 were further inhibiting factors. (These interviews are consonant with the findings on managers in Chapter 8.)

The probability of a long process of interaction with public health and other regulatory agencies, particularly in the absence of standards for drinking reclaimed water, increased reluctance to propose planned potable re-use. Most managers also felt public attitudes would be a problem. They all agreed that some people would support re-use but that this support would certainly not be unanimous. Public education could help considerably, they argued, but one could still expect opposition. In the one site where reclamation for drinking was considered (but not actually proposed), some members of the public, upon hearing of it, opposed it. The data presented in Chapter 8 support these managerial contentions. The general managerial assessment was that re-use was not completely consistent with the water-quality goal.[18]

Managers perceived some of the same problems, although as somewhat less important, with body-contact recreation (specifically swimming). Here again some managers viewed viruses as a problem. Even the successful Santee experience could not claim to have solved this problem, because of place-to-place variations in operations, input water quality, altitude, etc.[19] Santee provided many lessons, but each local situation called for certain modifications. Swimming in lakes filled with reclaimed water did not occur at the time of interviewing in the study sites. Santee and Lake Tahoe (Indian Creek Reservoir) barred swimming because of beach inadequacy and insurance problems. Antelope Valley, then not constructed, did not plan for swimming, nor did Canyon Lakes (Phase I) in Lubbock (street runoff problems rather than the quality of the reclaimed water were cited as the inhibitor).

The water managers in the study sites were more convinced of the easier acceptance of reclaimed water for body-contact recreation than for drinking. A number of factors help to explain this. Those who may object to using reclaimed water for recreation need not participate. They do not have this option with general municipal use, or may have to go to extra trouble and expense (for example, bottled water) to avoid using it. There is precedent for recreation in reclaimed water; there is less confidence in the precedents for drinking. The easier acceptance of recreation is also consistent with the use acceptance hierarchy identified in the public attitude survey and discussed in Chapter 7.

The differences between water and wastewater personnel and the nature of their linkage are relevant to a discussion of goals. Many of the technical managers interviewed were concerned with wastewater. These people are closer to wastewater treatment technology and to the quality of the effluent and are generally more likely to be strong supporters than water-supply personnel. In Santee and Colorado Springs the wastewater people were the major supporters of re-use. In San Angelo the wastewater reclamation champion was a water superintendent, but he had previously been in a wastewater department. Of course, wastewater personnel do have more incentive to favor re-use, in that it would provide more active roles for them. A client seeking wastewater for a given use is more likely to contact the manager of the wastewater system. The wastewater manager is also more likely than the water department manager to search for a use for this effluent.

The nature of the administrative links among departments can also be a factor. In Lubbock the water department was not really involved in the Canyon Lakes scheme; the City Planning Department dealt primarily with the wastewater unit. In San Angelo the wastewater personnel did not get involved to any extent. In Santee the organization was a water and sewer district directed by the same man; the joining of these functions eased desired implementation. In Colorado Springs the water and sewer departments were separate divisions of a municipal utilities system, but coordination was adequate. The reclaimed wastewater was initially handled by the sewer division, then management was transferred to the water department, and then subsequently returned to the sewer division. This was not a coordination problem but, rather, reflected a preference, commonly held, that the wastewater department be the purveyor.

It seems that water-supply personnel will be more resistant than wastewater personnel to direct potable re-use, and that this resistance will increase if communication, coordination, or experiences between the water or wastewater departments are poor. Water-supply personnel may be more likely to have stronger attitudes about perceived purity of the raw water source and more fears about contamination.

As noted earlier, there was general agreement among water managers (both water and wastewater) on cost and quantity considerations. And when reclaimed water is suggested for lower-order uses, the disagreement surrounding the quality goals lessens. Managers see the use of reclaimed water for lower-order uses as contributing to its eventual acceptance by the public for high-order purposes. Experience with reclaimed water (so long as it was favorable), the managers felt, would predispose the public toward greater acceptance. Such experience would also increase the degree of public and professional trust in the system and might thus place in a different context the inhibitions about high-order re-use.

The recognition of plural water-supply systems, and the water-quality hierarchy-of-use tradeoffs, is an indication of a tendency toward efficiency in the system. Acting on the opportunities presented by this hierarchy could mean employing re-use now; and in so doing one would not violate any public management goals. Although it would be easier with recreational use, employing reclaimed water for potable use would be perceived as challenging if not violating, the quality goal—in many ways the one most highly regarded, and, of course, the most closely regulated. In any event, it would probably not be necessary.

20. Gore, p. 77.

21. I do not wish to overstate the clarity of the goal-matching process. Clear and consensual goals can be assumed in the "isolation of the abstract" but "in reality goals are always surrounded by a thick, sticky coating of ambiguity" (Gore, p. 37). But some goal-matching does occur. See Gore, p. 37, and Herbert Simon, "On the Concept of Organizational Goal," *Administrative Science Quarterly*, 9 (1964), 1–22.

22. Peter Clark and James Q. Wilson, "Incentive Systems: A Theory of Organizations," *Administrative Science Quarterly*, 6 (1961), 157.

The use of reclaimed water for lower-order purposes may contribute to its eventual acceptance for higher-order uses. Even more importantly, however, it would release potable water for drinking. In this context predictions of planned, direct-potable use within ten to fifteen years seem optimistic—at least in the United States.

*International organizational goals (I-4)*. "When mission conception and internal goals overlap," Gore argues, "an organization may take a posture that mission commitments are all that count. Since this is seldom true, a second projection is needed."[20] This focuses on what the organization's members seek. The visibility of goal-matching will be higher on public management goals because these are the terms in which rationalization will primarily be couched. A matching process with internal goals does, however, occur, even though it is more informal.[21]

Four general goals will be considered:

(1) autonomy
(2) personnel stability
(3) client satisfaction
(4) professionalization.

These points, clearly not equally important to all personnel in the organization, are likely to be important to people in higher positions of the organization and therefore essential to the innovation.

Autonomy is generally highly valued by managers: "By autonomy we refer to the extent to which an organization possesses a distinctive area of competence, a clearly demarcated clientele or membership, and undisputed jurisdiction over a function, service, goal, issue, or cause. Organizations seek to make their environment stable and certain and to remove threats to their identities."[22] Managers usually wish to avoid regulation and the reduction of their area of administrative discretion. This constriction of discretionary behavior may come about through interest group antagonism, through conflict with another component of the system (for example, the council or city manager), or through demands imposed by a suprasystem (for example, the Department of Public Health regulation).

Wastewater reclamation and re-use, to the degree that it would involve such contriction of discretion, would seem to detract from the autonomy goal. This observation requires serious qualification, however, for industrial and agricultural re-use probably would not involve significant regulation. Regulation of the three types mentioned, but especially from the public health officials would, of course, enter into the applications of reclaimed water for higher uses, but in these cases the regulation would probably be welcome. No water manager or political official would presently attempt to implement a program of this type without Public Health Department approval, although the opportunity did present itself in Chanute (see Chapter 2). Professional approval, by the public health officials, with its attendant political implications, would more than compensate for any reduction in autonomy.

Certainly San Angelo would not have proceeded without public health approval. In Santee the prolonged interaction between the champions of innovation and the local health officer, a highly respected man, may have contributed substantially to public acceptance. The approval, after such long debate and delays, may have meant more to the public. The interaction also served to spread the political risk of, and reinforce political and organizational confidence in, the re-use innovation. In this context, then, the reduction in autonomy does not reduce the prospects of the re-use innovation, and, as stated earlier, poses no problem for lower-order uses.

Wastewater reclamation may also have some implications for the stability of departmental personnel (for example, job security, status positions). Among the factors affecting the rate of adoption of an innovation is the amount of change required in the system in which the innovation occurs.[23] One such change is the rearrangement of personnel that innovation requires. People who see their positions as being adversely affected by an innovation are not likely to urge its adoption.

One can argue that the implementation of wastewater reclamation for higher-order uses implies and demands a shift toward better-trained water managers and/or more extensive use of consultants and governmental technical personnel. Water managers in the study sites were generally competent and adequately trained. In most sites federal and state funds were being used to conduct experiments on wastewater treatment processes. In fact, it was largely through the efforts of these competent water managers that water re-use was practiced or proposed at many of these sites. In cases where the water and/or wastewater departments are staffed with personnel of inadequate training and experience, the threat to job security, position within the organization, or self-image (through encounters with poorly understood technical situations) implied by reclamation may involve attempts to block its adoption.

Inadequate training of water managers, then, militates against the re-use innovation by (1) making its initiation less likely (the earlier statements about communication nets, etc. apply here), (2) making interdepartmental opposition and conflict more likely, and (3) decreasing the confidence of other components of the water-management system in the operations component. These problems increase with an increasing order of use.

An important internal goal is client satisfaction. The general public is a client of the water-management system, but there are other, less diffuse users whose desires the organization seeks to meet. They include large commercial and industrial users and other municipal departments.

In a sense the municipal political system is also a client. If community growth is to continue, demands are made by such groups as the Chamber of Commerce, planning departments, and city council upon the water-management system to meet these water-resouce requirements. More than revenue is earned if these are met; the good will accrued can be tapped when requesting budget or staff increases, and personal prestige is also enhanced. This good will also allows for more administrative discretion for the water manager, as the city council, for example, feels less compelled to intervene in the department's affairs. A satisfactory performance also enhances the water manager's public image, at least among citizens who know who he is.

23. A number of innovation attributes are discussed in Gerald Zaltman and Nan Lin, "On the Nature of Innovations," *American Behavioral Scientist*, 14 (1971), 651–674.

24. See Francine F. Rabinovitz, *City Politics and Planning* (New York, Atherton, 1969) p. 133.

25. This image appeared strong in Lake Tahoe, Santee, and Lubbock. In San Angelo there was some enthusiasm, though mixed, about the prospect of being a leader in the field of re-use. There are also financial benefits (research grants, subsidization of facilities, personnel salaries) in being a demonstration or model site. That was certainly true at Santee and is another support for the "innovating organization" image.

26. In the areas of water-supply adequacy and public health, reliability becomes particularly important. Officials strive to avoid system failure even if it leads to economically inefficient behavior (for example, source expansion well in front of need). See Russell, Arey, and Kates, p. 195.

In Colorado Springs the municipal political system supported supply augmentation through re-use, as did such groups as the Chamber of Commerce. In Lubbock the clients for the Canyon Lakes Projects included the Planning Department, whose interest was innovative planning, and the City Council, who saw it as a way to enhance land values and as an enticement to industrial expansion. In Santee the clients were local and regional recreationists, a local water user, and the major land developer. Antelope Valley clients appear to be similar. Satisfaction of these client interests was a central element in each innovation process.

Although meeting local goals is essential to water managers, professional peers constitute another important reference group, and professionalism as a source of norms and prestige can have a considerable effect upon the reclamation and re-use decision. This effect would vary, of course, with the degree of professionalization. In cases where professional identity is low, "the consequences of violations of professional norms are likely to be slight, and the intensity with which they are adopted may be decreased accordingly."[24]

In the study sites professional identity was quite high. Information about norms as well as techniques flows through a complex web of communications. Of this information at least three areas are of direct concern—the norms on (1) high information levels, (2) innovation, and (3) efficiency and reliability.

Many of the managers interviewed emphasized the importance of their own high levels of information as well as the desirability of finding techniques to increase information levels among other managers. Virtually all managers share a norm on being well informed—as much in the vein of "education for its own sake" as for the obvious value of information in expanding choices and reducing the uncertainty associated with them. High levels of awareness constitute a source of personal prestige as well as of value in professional associations. Adherence to this norm by water managers enhances prospects for water re-use.

These prospects also benefit from a norm on innovation. An innovating organization can project a certain image (accurate or otherwise) of itself as creative, dynamic, competent, aware, etc.,[25] and thus secure the desired esteem of professional associates. This is not to say, however, that organizations innovate merely for the sake of innovation, even though some may. Norms also affect the direction and the significant attributes of the innovation. Economic efficiency and reliability are two of these attributes.

An earlier section (I-3) discusses the managerial perception of the efficiency and reliability of water re-use systems. These attributes (that is, characteristics of the innovation), particularly reliability, are very important to the managers.[26] There are local or intersystem pressures toward satisfying these goals, but professional norms and intra- (or supra-) system forces reinforce their significance. Managing an efficient and reliable innovation is an act that satisfies professional norms and is a source both of personal satisfaction and peer prestige. Where wastewater reclamation and re-use are consistent with these norms, managers are likely to favor them.

*Activity level (I-5)*. Through the goal-matching process, the organization (at this point, the technical component) has already established the reclamation and re-use opportunity as an object of legitimate concern. In the determination of the activity level (that is, which level of use to employ), the order of use that may have been implicit and vague in the goal matching now must become explicit and concrete. This is achieved by the interaction of the innovation champions with other components—the political system, suprasystem actors, and elements of the system's environment.

The hierarchy-of-use formulation so important to the goal-matching process is equally central here. The extent of the decision field will vary with the order of use (the higher uses will demand more extensive networks of involvement and more intense interaction within this network). A decision to supply a power plant implies a certain limited decision involving water managers, a lightly involved political component, and some system clients. On the other hand, a decision to augment the municipal potable supply with reclaimed water encompasses the widest network possible.

This section on activity levels makes some general statements on the interdependencies of an organization in an environment. Such activity levels relate primarily to the organization's latitude for independent action. The section outlines the progressively more formal process of environmental survey and illustrates the process of determining the activity level in some study sites. The emphasis here is on the technical component in this early stage of the innovation; the process is replicated with different considerations at later stages.

Strong interdependence often characterizes the interaction process between an organization and its environment, and this is particularly true if the association has been a long one. Newly formed organizations may start with a relatively independent set of objectives, but operationalization soon enmeshes them with their environment[27] and radically affects subsequent goal formation.[28] This interaction concretizes into a set of mutual commitments between the organization and its clients (groups in the environment which have a stake in the organization's performance).

These commitments frequently comprise a set of formal goals and an equally important set of informal objectives. They can "become terms of an unwritten contract, so that anyone who tampers with the sanctioned interpretation of goals will feel the sting of social control."[29] These costs, Gore points out, may appear as stresses, shifts in alliances, clientele irritation, budget curtailments, or revisions in statutes. Costs and benefits of a specific use of reclaimed water are projected for these considerations, as well as for economic and technological factors.

The degree of constraint on the organization's activities is directly related to the extent of structuring in the process of environmental interaction. Commitment of the organization's output is one example of this; in the wastewater re-use context this may mean competing claims for the effluent. In Lubbock the land-disposal terminus is an irrigator who also gains some control of the water by contract. This commitment could constrain the city's behavior in the use of effluent for purposes of industrial cooling and recrea-

27. At least one author has integrated this into an "organizational life cycle" (his example is a regulatory agency). See Murray Edelman, *The Symbolic Uses of Politics* (Urbana, University of Illinois Press, 1964).

28. This is noted by, among others, James Thompson and William McEwen, "Organizational Goals and Environment: Goal Setting as an Interaction Process," *American Sociological Review*, 23 (1958), 28–29.

29. Gore, p. 76.

30. See Anastasia Van Burkalow, "The Geography of New York City's Water Supply: A Study of Interactions," *Geographical Review*, 49 (1959), 369–386. For a treatment of this point regarding recreation on water supply reservoirs, see Baumann, *Recreational Use*. For an abstracted argument, see Baumann, "Perception and Public Policy in the Recreational Use of Domestic Water Supply Reservoirs," *Water Resources Research*, 5 (1969), 543–555.

31. The superintendent of the Water Department at San Angelo "had long been a strong proponent of re-use," and had frequently discussed it with others in the system.

32. For an interesting treatment of the "search" notion in environmental scanning, see Herbert Simon, "The Decision Maker as Innovator," in *Concepts and Issues in Administrative Behavior*, ed. Sidney Malich and Edward Van Niss (Englewood Cliffs, N.J., Prentice Hall, 1962) pp. 66–69.

tional lakes. Colorado Springs must release a certain amount of effluent to meet the established rights of downstream users. This is to suggest not that an organization is a complete prisoner to these commitments, but rather that such obligations can severely constrain its behavior.

Although a survey for application opportunities is a useful part of the innovation process, an earlier and more important review must be conducted—namely, a survey of the environmental consequences of using the effluent in a specific way, and how this use might affect the present set of commitments. The search process is usually very informal and tentative and precedes even the commissioning of a consultant's report. It is a delicate point in the innovation process. It is also a point at which horseback judgments by water managers are particularly important. A number of authors have noted the impact of these judgments on other water-policy situations.[30]

The rationale for this survey is to assess the conditions of the environment with respect to re-use opportunities. Even though this process has been occurring frequently during the previous innovation activity,[31] it can be treated here more distinctly. A range of things may trigger it: a possible failure of water-supply alternatives, as at San Angelo, a commitment to use an existing facility (new sewage treatment plant), or an attempt to avoid a certain outcome (joining the San Diego Metropolitan Sewer District), as at Santee. The degree of testing and its tentativeness will vary with the order of use. Proposals for lower-order uses could be made easily, but for higher-order uses (particularly drinking) a more cautious approach would be appropriate.

Innovation supporters may pursue somewhat different but not exclusive approaches in surveying each of two environments—internal and external.[32] The internal environment, which includes the other components of the water-management system, must explore its varying perceptions of need, cost factors, uncertainty, and willingness to tolerate opposition. Appropriate approaches in this context are small group meetings and conversations between individuals. The second stage (the political component) of the model provides more extensive coverage of these points.

The external environment comprises the general public, certain interest groups, and, in a more special sense, suprasystem agencies and the system or system's clients. Although public information techniques are suitable for the internal environment, informal contacts with the external are usually more appropriate. These techniques are exploratory and are intended more for generating feedback than for providing information. Indeed a formal public information campaign by the water-management system implies a good deal about the level of agreement in the system. Once again, these techniques are not mutually exclusive. In Lubbock the Planning Department gave a slide presentation on the Canyon Lakes Project to the City Council, and in San Angelo consideration of reclamation did not get beyond the small-group level in dealing with the external environment. Stage III of this model treats aspects of the system's interaction with the external environment.

Reports on the major individual sites pinpoint various techniques of obtaining information on the state of the environments. Each site, however, usually enlisted some combination of three general procedures: (1) feedback from public information efforts; (2) feedback from a certain activity; and (3) the presentation of client demands.

Innovation managers cited public speeches at meetings of community groups, small "coffee klatsches," and feedback from these meetings as key elements of their program from beginning to end. Though public reaction was important throughout, it was a particularly valuable source of information in the early stages of innovation formulation and justification. In later stages these public encounters were more strongly oriented toward persuasion than exploration. This early stage includes, of course, communication with members of the internal system. In fact, response from this segment is more important than public response, for it is the internal-system managers who have the formal authority and control over resources needed to carry on the program.

It is possible to elicit public feedback for more sensitive issues without specifying the context. In San Angelo the mere consideration by the City Commission of direct potable re-use activated the beginnings of a possible community conflict. Yet earlier, and even during the process, the water superintendent was able to appear at public gatherings, drink reclaimed water, and endorse re-use without identifying the San Angelo situation. In this way, he was able to obtain some information on public reactions which positively affected his assessment of re-use possibilities.

Another procedure is to initiate some kind of re-use and wait for the public reaction. Santee landscaped some early lakes and added some facilities, but officials prohibited public entry and fenced off the area. The consequent public response abetted the innovation champions' justification for proceeding. The adoption of an innovation in one place can eliminate much of this preliminary exploration and need to convince. Given the widely acknowledged success of the Santee model, the Canyon Lakes in Lubbock and the Antelope Valley Project in Lancaster were easier.

Client demands constitute a rather narrow source of information, because clients come with a fairly specific set of demands which they hope the water-management system will meet. Although these demands are often narrowly based, usually in industry or agriculture, they can play an important part in a re-use program. First, it is possible to dispense with them fairly rapidly and, as mentioned earlier, satisfy a number of system goals. Secondly, and more important, they establish in the community the principle of beneficial use of effluent from which late re-use activities can profit.

Even this informal environmental survey provides managers with valuable insights regarding the pace at which to proceed, potential problem issues, and likely sources of support or opposition in the environment. Chapter 7 shows that an inverse relationship holds between order of use and levels of public support. This relationship, which emerges from the pattern of public responses, is an important factor in managerial perceptions. The gap in public acceptance regarding body contact and particularly ingestion indicates that a threshold exists among managers also. Even the most enthusiastic advocates of wastewater reclamation and re-use were cautious about body contact and very reluctant to recommend ingestion. Related to the hazard potential is this greater likelihood of public opposition, something nearly all water managers choose to avoid. Chapter 8 provides a detailed discussion of similar concerns among public health officials and consulting engineers.

For the lower-order uses the environment in the study sites was initially very favorable. The replacement of potable with reclaimed water for a range of uses including power-plant cooling (a municipal utility) was very well received in Colorado Springs. These primarily administrative decisions between the clients and the water-management system involved no conflict and served, in fact, to enhance the public image and increase good will among the management-system components.

A similar situation regarding power-plant use of effluent prevailed in Lubbock. The Canyon Lakes involved a few more issues, but because the lakes are primarily ornamental and the intended use did not involve body contact, the problems relating specifically to the use of reclaimed water were minimal. Initial reaction of the internal environment to planning-department proposals was generally favorable and sometimes enthusiastic. The multifaceted Canyon Lakes Project (see Chapter 3) was a campaign plank for some enterprising politicians, who saw it as attracting considerable political support.

Higher-order uses are more likely to encounter an internal environment of ambivalence or hostility. Champions of innovation must then decide whether to proceed and try to push it through, try to convince certain key actors in order to tip the balance to a favorable point, or drop the innovation effort. Since this book deals with communities that are currently interested in or have implemented wastewater re-use, it does not examine systems hostile to the idea (that is, places where the idea was either immediately dismissed or firmly rejected). "Hostile" here means organized immediate resistance by groups in the environment.

An ambivalent environment does not preclude hostility, but rather indicates either mild opposition or stong opposition offset by equal or greater support for the innovation idea. Hence consideration of the innovation does continue. The City Commission of San Angelo, divided in its view of wastewater reclamation for drinking, nonetheless interacted with state and federal officials and commissioned studies by consultants. Consideration of direct potable re-use ended only with the rains. A central problem was the high level of use intended; had the system's use structure differed—for example, if large users could have accepted effluent and released potable water,—adoption of a re-use system probably would have abated the water-supply problem.

The environment in Santee was much more favorable but still ambivalent. Again this was largely due to the high level of intended use (swimming); lower uses presented no problems. Innovation supporters interacting with the Department of Public Health substantiated the approval by a majority of the Board of Commissioners and helped convince the opposition. Most important here was the interdependence of the actors in the internal and external environments.

If the initial reactions of the internal and/or external environment are negative, the innovation champions (herein the technical component) may decide to table the innovation process and add the experience to its fund of environmental information, where it will affect subsequent opportunities to propose re-use. Or the technical component may alter its level of intended use and renew the surveying process.

Nearly all water-management decisions that involve contracts or funding are submitted for council approval. These decisions include those termed "administrative" (for example, industrial supply of effluent), but usually involve only formalization. For higher order uses, this interaction is, as indicated, more than mere formalization.

The political component of the water-management system and certain prominent actors (for example, the Department of Public Health) receive an assessment of the reclamation and re-use situation from certain points of view (economic, technological, health hazards). Their acceptance even of these points will depend on many things, such as the water-management system's history and the prevailing attitude toward the water managers. In any case, these actors will then review the proposal from their own points of view. The search for the meanings of re-use in these terms must proceed before the system can act as a unit. The elements of this process form the second stage of the model.

## The Political Component (II)

The central actors in the political component of the water-management system are the elected officials—mayor, councilmen, water commissioners. Such formal groups as citizen's advisory boards could be included, but they were not important in the study sites. Other political/administrative bodies, such as a state legislature or a public health department, are more usefully treated as elements of the environment in which the system must act. This is also true of clients such as large water users and interest groups.

Because the ultimate power regarding large municipal expenditures rests with the political component and because its behavior can legitimize a program, its role in the wastewater reclamation decision is critical. The water re-use proposal can here be translated into a political program. The organizational relevance of the re-use proposal is again established and some of its implications outlined and codified. Certain steps, such as commissioning a consultant's report, may be taken easily and others only with great debate. Much of this debate takes place in this second stage.

The water re-use proposal, formed from a certain perspective of costs and benefits, encounters other viewpoints in Stage II. It comes from a technical component responsible for a few functions to a political component responsible for many. The re-use proposal is placed in a system that has many competing demands for its resources; a system whose members have their own objectives and images, which may be inconsistent with the technical perspective; a system that may manage ambiguity and uncertainty differently than would the technical component. Elements of the re-use proposal may be replaced or others may be added, but it is usually altered, and the alterations may be severe if the technical component's evaluation of the proper activity level is faulty.

33. The example of Antelope Valley, California, and to some extent Lubbock, is instructive in this regard. In part the greater role of nontechnical people in these innovations is related to the greater degree of confidence in reclaimed wastewater lakes (particularly for non-swimming) because of the Santee experience. In effect, it involved less pioneering.

34. Rogers suggests that changes in stages may mean change in the perceived characteristics of the innovation and differences in adopter characteristics. See Everett M. Rogers, *Diffusion of Innovations* (New York, Free Press, 1962).

*The political component and the information network (II-1)*. How did the idea of the wastewater reclamation enter and gain initial support in the political component? As indicated, actors in the technical component originally championed innovation and will probably continue to do so in the pioneering stages of the innovation.[33] Wastewater personnel were usually the movers, although in Lubbock the City Planning Department was the major actor. In Santee the relevant political unit (no formal "city government" existed), the County Water District, and the technical function were so intertwined and closely coordinated that this distinction is not very clear. In most cities, however, the distinction is fairly clear and the technical component will probably be the focus of innovation activity *during the pioneering stage*.[34]

The reaction of the political component to a proposal advanced by the technical unit will be affected by the nature of the relationship between them. Although there may often be exceptions, the relationships between the technical and political components are usually adequate. Consistent failure to perform the technical services to the satisfaction of the political component would normally lead to a change in the relationship (for example, firing a department head), so that the relationships that survive are usually those in which mutual expectations have been met.

Confidence in the competence of the technical unit can vary, even if the political component is satisfied with its previous performance, and this will be particularly true for higher-order uses of reclaimed water. The result of this is not merely commissioning an engineering consultant's report. This usually occurs in any case ranging from normal supply expansion to consideration of direct-potable reuse. The change may take the form of a reassessment of the technical component's competence to function adequately on this new level and subject to the necessary degree of monitoring by the suprasystem of departmental activities. In Colorado Springs re-use proceeded with no change in the situation. In Lubbock there was somewhat more activity, but because of the limited scope of the soon-to-be-implemented facilities, little suprasystem regulation was involved. In Santee this was not true, and monitoring was considerable. In San Angelo close monitoring would have been imperative, and a possible realignment to higher, permanent staff technical capacity may have ensued.

The political actors in the study sites were aware of examples of agricultural and industrial uses of reclaimed water and were enthusiastic about possible applications in their area. Although they knew less about reclamation projects for recreation and drinking (in part because the examples are fewer), their information about them was greater. The levels of information, not surprisingly, tended to match the intended use. In San Angelo, where drinking was considered, information about direct-potable re-use was higher. In Lubbock the managers tended to be best informed on recreational lakes, most prominently those at Tucson and Santee. In contrast to the national examples cited for higher-order uses, agricultural and industrial applications were usually regional.

Technical and safety dimensions of the innovations present options and constraints to the political managers, and these are matched against an image, however fluid or rigid, of what this innovation may mean to their clients,

the general public and/or certain interest groups. Surely this process occurred in San Angelo, where the possibility of community conflict was discussed, and in Lubbock, where the obvious political and economic benefits
of the Canyon Lakes scheme were identified.

The political actor and the technical actor each occupy in an information
network a respective node which can, depending upon its occupant, be large
or small, open or closed. At least two characteristics of the occupant which
affect this network are professionalism and ambition. The professional can
be a city manager or a politician.

The city manager, partly because of his tenure situation and professional
orientation, exists in an information network that is largely extralocal.[35] Conferences, journals, and professional background tend to expose him to
newer ideas, programs, and techniques. The presence in the system of
these professionals at least increases the probability of contact with wastewater re-use. In contrast, the elected official with a local orientation is less
likely to have these links.

The differing orientations of elected officials with political ambitions and
those with short-term goals can also considerably affect exposure to information on wastewater re-use. Presumably an official who is developing his
political future will form wider links and become exposed to more contacts
with wastewater reclamation projects, often on an important personal or
small-group basis. A mayor with wide links in a water-short region is much
more likely to encounter water re-use examples than a mayor with a purely
local orientation. There may also be differences in his susceptibility to local
influences, sensitivity to conflict potential, and attitudes toward the innovative image.

Political managers share a similar orientation to the type of information
they desire. Some may care only for the political aspects of the re-use experience and leave the technical aspects to the technical component.
Others may care for both but place their main emphasis on the political dimensions. For example, in San Angelo political managers were aware that
Windhoek had achieved technical implementation but lamented the lack of
information on public reaction. Although the virus and weed management
problems at Santee were objects of concern, the political managers in Lubbock expressed special interest in public reaction, information presentation
strategies, and their relevance to the Lubbock context.[36]

This process of environmental survey will receive more explicit treatment
later in this chapter (II-5), but some observations are pertinent here. Perhaps
the most important is that the amount of energy expended on environmental
assessment increased with higher levels of use. This was also true of the
degree of caution in the survey.

Once they had reached agreement, Lubbock officials merely announced
the provision of effluent for the power plant. This use of effluent may even
have contributed to the acceptance of Canyon Lakes. The Canyon Lakes
Project was formally proposed to the Council after considerable effort to answer a range of questions and to provide an elaborate rationale for the venture. The same presentation that the City Council received was then widely
shown to various groups in the community. In Santee, where swimming was

35. See Ronald
Loveridge, *City Mana-
gers in Legislative Poli-
tics* (Indianapolis,
Bobbs Merrill, 1971),
pp. 44–47.

36. For a discussion of
how certain publics can
affect a water man-
ager's latitude for ac-
tion, see Charles Garri-
son and Duane Hill,
"The Dynamics of Pub-
lic Roles in the Selec-
tion of Revenue
Sources in Local Water
Administration," *Water
Resources Research*, 3
(1967) 949–962.

37. Kasperson, "En-
vironmental Stress."

planned and where the idea was regarded as very new, the managers de-
voted considerable effort to community and small-group discussions and
enacted an incremental schedule for implementation. In San Angelo, where
drinking was considered, the water manager spoke on water re-use to
groups and sought informal opinion but did not propose the idea of direct-
potable re-use. Discussion of this alternative was confined to the City Commission.

The environmental survey includes the formal and informal techniques of
obtaining general feedback on a given action. A more deliberate survey employs probes by which one identifies potential allies and opponents. Another
and very important means of environmental survey is an analysis of client
demands. In fact, this method may assure support for an innovation. Demonstration of a need (particularly if the technical component's clients are also
the political component's clients) increases the likelihood that potential innovation will receive attention.

*Water-resources situation (II-2).* Two previous sections have outlined the
water-resources situation in the study communities. Chapter 3 provides a
more detailed picture of each water problem, and one previous section (I-5)
in this chapter includes some generalizations about the water-resource situations in the communities. The present section concentrates on the aspects
more relevant to the political component. Of prime concern are the differences in perspective between the technical and political components and the
relationships between them which affect wastewater re-use prospects.

Although the technical people are responsible for one or two functions (for
example, water supply or wastewater), the political management system is
responsible for many and therefore must be responsive to many clients. The
two components are not likely to regard a problem as being of equal importance, or at least of equal urgency. A technical manager may early be convinced of the need for supply expansion, but one should not expect that the
political component will act immediately on his recommendations. As Kasperson puts it, if somewhat too strongly: "This illustrates a simple political
fact-of-life about a declining municipal water supply system—it does not
generate as much public concern as do many other municipal problems.
Until the supply becomes so depleted as to necessitate rigid use restrictions or to impair seriously quality of water, an unfavorable WU/SY (water
use/safe yield) ratio is a worrywart for bored engineers and unoccupied
water superintendents."[37]

This supply and quality problem did occur in San Angelo; political managers were very much concerned and the problem captured the attention of
the whole water-management system. This was due in part to the shortage,
in part to the high level of use under consideration. When the whole system
is involved, and particularly when the question is potable use, there are more
preferences to be coordinated. In San Angelo certain political actors were
less convinced than the Water Department that reclamation was safe and
acceptable. For a number of reasons, this among them, San Angelo did not
adopt water re-use.

Although one can conjecture that the political component will be more sensitive to the question of expected public reaction, the expansion of the "decision field"[38] does not necessarily put wastewater re-use at a disadvantage. In Lubbock, the expansion of the decision field served to aid the re-use prospect. The same occurred at Santee even though implementation proceeded incrementally. In the case of fluoridation the expansion of the decision field generally had a negative impact on its prospects for adoption.[39] This is not necessarily the case with wastewater re-use, but the probability of similar results may vary with the level of use.

The point raised earlier about unequal perception of the importance of water-resource questions by the technical and political component has two implications. One relates to urgency and the other to resource allocation. Urgency affects the amount or timing of attention that water questions receive. Delay does not automatically lead to rejection of re-use; it can, however, affect its prospects. There are points of choice. If expansion of wastewater treatment facilities is imminent and the potential of re-use receives insufficient attention, the increment in treatment quality may not be approved and the prospects for re-use thereby reduced. In Lubbock re-use proponents persuaded the City Council to include the Canyon Lakes Scheme in a tornado bond issue. Otherwise, convincing the City Council to proceed with another referendum would have been difficult. Similar considerations are relevant to the allocation of other resources (finances, personnel) to reclamation and re-use.

These questions of specific allocation exist within a larger framework which determines the general orientation of the political component toward water questions. The study sites were in dry areas, and the policy orientation toward supply expansion was particularly strong. Attention to water-supply questions, therefore, was quite high. The technical component shared the orientation of the political component toward supply expansion. Reclamation for uses other than drinking was well received by the political components in these sites, even when the need was questionable (although the growth orientation of some of the sites could justify its need through its projections).

The orientation toward expansion of water supply reflects the water-management system's perception of its own needs. In each site an additional set of needs imposed by the state were also being satisfied. The political component welcomed the water-supply benefits which accrued from increased water pollution control. Chapter 3 describes the circumstances, most of which were under pressure from state standards, surrounding wastewater treatment increments. This trend toward greater restrictions on waste discharges enhances the prospects of wastewater re-use, and in the study sites the political component responded favorably to the benefits of re-use identified by and with the technical component.

There is strong similarity between the goals of the technical and political components. Each component supports the quantity goal, particularly when the municipal policy is to pursue high growth. But even where there is not a pronounced emphasis on growth, the aversion to perceived shortage is strong, and restrictions are reluctantly imposed. This is not to advocate crisis as a management policy but only to suggest that willingness to employ use restrictions or peak-pricing schedules can provide management flexibility

38. By "decision field" I mean the number of actors who participate in a decision and the range of considerations which must be dealt with. This automatically expands with the increasing level of use. The actors are here considered to be participants, although each can represent one of a number of points of view and, therefore, articulate the preferences of nonparticipants.

39. Great emphasis is placed on strong executive leadership in fluoridation adoptions. Donald B. Rosenthal and Robert L. Crain, "Structure and Values in Local Political Systems: The Case of Fluoridation Decisions," *Journal of Politics*, 28 (1966), 169–196.

40. See the discussion in Russell, Arey, and Kates.

41. Of approximately 900 respondents, 37.5 percent identified abundance as the most important characteristic of new water sources, 25.3 percent cited quality, and only 14.6 percent asserted cost. William Bruvold, "Public Opinion and Knowledge Concerning New Water Sources in California," *Water Resources Research*, 8 (1972), 1145–51. Reclaimed water received a low performance rating, but to what extent this was because of low levels of knowledge or a negative attitude is not clear. For an early piece, however, see also W. Bruvold and Paul Ward, "Public Attitudes toward Uses of Reclaimed Wastewater," *Water and Sewage Works*, 117 (1970), 120–122.

42. Baumann, *Recreational Use*.

43. Robert Crain, Elihu Katz, and Donald B. Rosenthal, *The Politics of Community Conflict: The Fluoridation Decision* (Indianapolis, Bobbs Merrill, 1969).

44. See H. G. Frederickson and H. Magnas, "Comparing Attitudes toward Water Pollution in Syracuse," *Water Resources Research*, 4 (1968), 877–889. See also Edmund Costantini and Kenneth Hanf, *Environmental Concern and Political Elites: A Study of Perception, Background, and Attitudes*, University of California at Davis, Institute of Governmental Affairs, Research Reports, 21 (Davis, California, 1971).

(see Chapter 1).[40] The identification by the public of quantity as the most important factor in consideration of new water sources suggests its support for this goal.[41] This finding relates to Southern California, which has been subject to hard campaigning on this issue, and it may vary from region to region. The evidence from the study sites suggests, however, that it holds for other areas as well.

The level of water quality desired may also vary regionally,[42] but higher quality, particularly hazard-reduction potential, was a major consideration in the higher-use sites: both the political and the technical components were concerned about it and anxious to avoid offending the public and the regulatory agencies. Although hazard concerns both components, they may in a political context act upon it differently. A technical definition of hazard involves judgment and frequently identifies safety as including some small possibility of hazard. A political component that is not unified may in a public conflict come to regard even this level as too high. Some fluoridation controversies illustrate the point.[43]

The cost of water fluctuates and is in a sense a translation of the quantity and quality goals. Water managers will pay heavily to obtain pure upland reservoir sources to satisfy perceived public preferences. They will also invest substantially to assure an adequate future supply for their community in competition with others. Consumers will use much water for nonessential uses at low prices. Indeed, cost may be a very flexible goal. Given a narrow choice and assuming sufficient quantity and quality, a water system will invest in a very expensive new source. Yet the political manager shares the technical manager's reluctance to use pricing as a vigorous management tool. Wastewater reclamation and re-use would meet the cost goal because the necessary task of waste treatment could provide part of the needed expense. The political component also viewed re-use as furthering this goal through the provision of low-cost water to special clients.

The Canyon Lakes Project satisfied certain criteria of the water-management system and, more importantly, contributed to a number of other goals the political component regarded as important. Promoted as a multipurpose project, it encompassed a wide range of goals and served such politically powerful clients as realtors and the business community. Similar if smaller-scale considerations were undoubtedly important at Santee and Antelope Valley. The projects contributed to the City Beautiful image and received the support of the more civic-minded organizations. Obviously the political component had much to gain by satisfying these clients.

Equally obvious is the potential contribution of re-use to the growth orientation of the political component. Actors clearly note the opportunities for expansion of a low-cost industrial supply. The implementation of re-use allows the political component to wave the ecological banner and to highlight its fastidiousness in water pollution control. This would appeal to a politically supportive segment of the population, while retaining some of the benefits of this "new, clean water" in the community.[44]

The willingness to engage in a political dispute becomes a primary consideration in re-use adoption. Faint hearts in the water-management system (both technical and political components) during the conflict contributed to a number of fluoridation failures. The conflict can involve groups in the general public, regulatory agencies, or both. Political components will differ in their willingness to risk conflict. If there are few alternatives, then the political component may have little choice; but then one would expect less conflict.

Other dimensions of the political component may affect its willingness to adopt direct potable re-use. The local government's conception of its proper function (the "caretaker," who provides services versus the "arbiter," who resolves interest group disputes) may be a factor.[45] Certainly a commitment to wide community participation in decision-making would affect the innovation process,[46] although it might not substantially affect the initial commitment to engage the situation. Whether the decision on re-use is positive or negative, the innovation at any level calls for a number of perspectives, each involving its own calculus of utility.[47] A position on the higher levels of use involves trade-offs on a number of points.

*Internal management goals (II-4)*. Norms in the political component are more elusive than those in the technical. Since there is relatively less professionalization (for example, engineers), there are fewer common specific traits (for example, emphasis on efficiency and innovation) whose importance for the re-use innovation one can identify. While the political system clearly has a complex set of norms affecting its behavior, the impact on re-use is unclear.

That individuals with their specific attitudes can greatly affect the outcomes of community decisions is clear. This means, of course, that the attitudes toward re-use (ingestion of wastes, etc.) among the individual actors in the political system can affect its prospects. One premise of this research is that water managers project their own attitudes upon the public and then deny serious consideration of re-use on the assumption that the public will not accept it. The findings of research on consulting engineers and public health officials (Chapter 8) lend credence to this position. These attitudes will also work on a more conscious political level as the intended level of re-use is debated.

The individual's attitude will have less effect if the situation is more structured, or if his position is strongly opposed by others. The impact can nevertheless be considerable. Kasperson characterizes the resolution of the Brockton, Massachusetts, water crisis as being "in large part . . . a tale of two mayors,"[48] The impact of individual attitudes had considerable influence in some fluoridation conflicts and in other water-resource decisions as well.[49]

Strong support by water managers is essential if re-use is to occur. Positive attitudes not only affect the vigor with which managers champion re-use but also, importantly, lead them to identify water-resource situations which can benefit from re-use.

To say that many organizational goals provide rationalizations for projects that have been undertaken rather than reasons for engaging in these activities is perhaps too cynical.[50] Although the goal of "wise water use" was frequently cited, client satisfaction may be no less important for managers. Many of the considerations (for example, levels of knowledge, the perceived water-resource situation, the public-management goals, and the internal

45. See Oliver Williams and Charles Adrian, *Four Cities: A Study in Comparative Policy Making* (Philadelphia, University of Pennsylvania Press, 1963).

46. Wide participation in the fluoridation decision generally led to its rejection. See Rosenthal and Crain, p. 196.

47. See Roland McKean, "Costs and Benefits from Different Viewpoints," in *Public Expenditures in the Urban Community*, ed. Howard Schaller (Baltimore: The Johns Hopkins Press for Resources for the Future, 1962).

48. For a description of drought in this context, see Kasperson, "Environmental Stress."

49. See Irving Howards and Edward Kaynor, *Institutional Patterns in Evolving Regional Programs for Water Resource Management* (Amherst, University of Massachusetts, Water Resources Research Center, 1971), p. 4. See also Russell, Arey, and Kates, p. 171.

50. Total emphasis on upland reservoirs for water supply may be subject to this. See David Rados, "Selection and Evaluation of Alternatives in Repetitive Decision-Making," *Administrative Science Quarterly*, 17 (1972), 196–204. Repetition of a decision situation provides opportunity and incentive both for rationalization of the decision process and development of a decision program that embodies the rationalization and reduces the cognitive load required (ibid., p. 196). A similar view is presented in

198       **David McCauley**

Cohen, March, and Olsen: "such organizations can be viewed for some purposes as collections of choices looking for problems, issues and feelings looking for decision situations in which they might be aired, solutions looking for issues to which they might be an answer and decision-makers looking for work." Michael Cohen, James March, and Johan Olsen, "A Garbage Can Model of Organizational Choice," *Administrative Science Quarterly*, 17 (1972), 1.

51. Even if not "water-starved." Projections go on, and demand goes up; the thirst for new sources seems unquenchable. See Robert Fleming and Harold Jobes, "Water Reuse: A Texas Necessity," *Journal of the Water Pollution Control Federation*, 41 (1969), 1564–69.

goals) mentioned in other sections come to focus in the selection of a particular activity level in a specific situation.

*Activity level (II-5)*. The Technical Component activity section (I-5) includes some general remarks on the meaning of activity level. The similarity in behavior, because of the need to coordinate activities, between the technical and political components should not be surprising. More detailed statements on the interaction process in each community occur in Chapter 3. This section will limit itself to a few generalizations about each site. Again in this section, the notion of hierarchy of use and its impact on the decision field forms an essential background.

In Colorado Springs the re-use innovation was for such lower-order uses as industrial water supply and irrigation of parks, golf courses, and institutional grounds. Although Colorado Springs has a large supply capacity, projected high growth figures indicate a future problem. The use of nonpotable water and the release of potable water for other uses were seen as part of a long-range planning program. The provision of reclaimed water for these lower-order uses contributes to some important goals. Growth is one; a city that has apparently enjoyed public endorsement of heavy expenditure for new supplies found a new and expandable source. All of the water-management system's goals (quantity, quality, cost) are met. Re-use has met the needs of other municipal departments (electric utilities and parks) and some private users. The city has been approached by a group interested in re-use to provide a recreational-ornamental lake in a dry area.

There was, of course, no adverse public reaction; it was, in fact, almost unanimously favorable, and enhanced the image of the Sanitation Department. Possible hazards from re-use were, because of the level, no problem. The various uses were achieved through simple agreements between the client and the Sanitation Department. Clients were satisfied, certain goals were advanced, and only basically administrative agreements were involved. Re-use at this level is easy and efficient.

In Lubbock, Texas, the Canyon Lakes Project became a political issue, but one for which there was nearly universal support. No pressure on water supply occurred (sources are adequate to 1990, when new sources are planned), but there is usually enthusiasm for new sources in the water-hungry region.[51] The provision of water for a new power plant was one aid to municipal growth; the Canyon Lakes, as a source of industrial water, would provide more. The political component applauded and was even more enthusiastic about the other benefits of the Canyon Lakes scheme. The provision of basically ornamental lakes certainly aided local political goals.

Since swimming was not planned, there were no objections to the Canyon Lakes scheme on health grounds. The project has satisfied the objectives of many groups—business interests, community groups, and city planners. The two-to-one acceptance rate on the bond issue indicates that the general public also approved.

In Lubbock the re-use innovation was accompanied by a major public promotion effort. Strong objections, which could practically have been based only on cost, were not forthcoming. The water-management system (with the Planning Department) advanced all of its goals and jeopardized none. Essentially, it proceeded on its objectives and had to make no major concessions to a hostile or ambivalent environment.

This was less true at Santee. Since the Santee County Water District was tied into Colorado River Water (and now the California Water Plan), local water insufficiency was not a factor. But the cost of this water was; one benefit of the Santee project was the provision of a lower-cost water for some of the clients. Another and more important benefit was the provision of the recreational and ornamental facilities. Early interaction between the water district manager and a client (a golf course owner and local developer) was important in initiating the lake project.

The checks on the speed and level of the re-use innovation were not internal constraints by the political component; imposed externally by the suprasystem, they arose because of perceived violation of the quality goal. The major constraints came from the County Department of Public Health, the watchdog of this goal.

The activity levels at Santee proceeded in essentially three stages: (1) opening the lakes for non-body contact recreation, (2) the "take home" fishing program, and (3) a swimming pool with reclaimed water. In each case the activity was established only after a conflict-cooperation interaction between the innovation's strong proponents and the Department of Public Health. The Public Health officials always voiced their objections on the basis of perceived hazard, some violation of the quality goal. In each case, after their objections were satisfied they cooperated in establishing the agreed-upon re-use application.

The recurrent conflict between the proponents and the regulators greatly affected other possible constraints on activity level. The District Board of Commissioners, the relevant political body, supported the innovation champions but usually conditioned final approval of a given increment of use upon the sanction of the regulators. It seems likely that public reaction was also affected. The image of the local health officer was one of "toughness but fairness," and his sanction, after a series of objections, may have seemed more valuable. But the innovation proponents at Santee did not leave this to chance, and they employed the fairly involved strategy of gaining public approval described in Chapter 7.

The impact of specific clients on the innovation was greater at the beginning. In the later stages the innovation was proceeding for its own purposes, and the clients became progressively more diffuse. Santee Lakes could have been a local and regional attraction even without continuous elaboration of the facilities. Research and experimentation had become primary goals, and the clients included a more widespread technical community.

In a sense the clients for reclaimed wastewater at Lake Tahoe were post hoc. The stimulus for high treatment levels and export from the Tahoe Basin came from a desire to Keep Tahoe Blue and an aversion to even a high-quality effluent, which was a potential threat to this extremely important objective. Indian Creek Lake has recreational potential but no facilities. Farmers are now using the effluent for supplemental irrigation, but there was not a strong organized effort at the time that the export arrangement was proposed to ensure the allocation of these supplies.

In San Angelo, Texas, the potential re-use clients were not presented with the choice. Only San Angelo considered re-use in an immediate water-supply context. Supply failure had created a drought or crisis condition. The city was in the process of expanding its wastewater treatment facilities, but effluent quality at the time of the crisis was inferior to that in the other sites. Had the effluent quality already been good and the increment been designed to bring it to an even higher level, the prospects for re-use would have been more favorable. Because the replacement of large potable uses with sewage effluent (thereby releasing fresh supplies) was not an option and because of the serious supply shortage, the level of use discussed by the City Commission was drinking. The client for this use was therefore the general public.

Whereas the hazard-related constraint was imposed externally at Santee, it was done internally in San Angelo. Because of the hazard potential and because of anticipated problems with public reaction, the City Commission did not actually propose direct potable re-use. When news of its discussion by City Commission chambers did reach the public, there was some initial adverse reaction.

*Suprasystem approval (II-6)*. Any number of suprasystem agencies (Santee had six) may be involved in any one re-use innovation, but the major actors are the various public health departments. Their role is pivotal because their approval must be considered essential to the successful prospects of re-use innovation.

In the study sites the importance of the public health agencies varied roughly with order of use. In Colorado Springs they played no role at all. In Lubbock, Texas, apparently because the proposed uses did not include body-contact recreation, the agencies did not affect the decision. In Lake Tahoe the reclaimed water was cleared for all body-contact recreation, but the high water quality reflected Alpine County's stringent standards and not the demands of public health officials. Were the effluent quality lower, interaction with health regulatory agencies might have been necessary.

In San Angelo, Texas, no significant interaction with the Public Health Department occurred because re-use had not actually been proposed. Otherwise such interaction would certainly have ensued and would have considerably affected the innovation procedure. Certainly, perceptions of potential public health agency response and public health standards had already been playing a continuous role in innovation formulation.

In Santee, California, involvement with the public health agencies was a central activity in the process of accepting and adopting wastewater reclamation. This interaction can perhaps be best termed a conflict-cooperation nexus, for while the public health officials resisted what they regarded as overenthusiastic moves on the part of the re-use champions, they cooperated in the research about and removal of the obstacles which they themselves had placed there.

Regrettably there were few examples of innovator-regulator interaction in the sites; their central role in higher-order innovations makes investigation of this question imperative. Statements about the standard-setting process, the impact of technical uncertainty about hazard potential on agency attitudes and behavior,[52] and variations in these factors by region and by need would be extremely helpful in assessing the potential for community adoption of wastewater re-use.

## The Community Component (III)

Up to this stage most of the innovation activity has been internal to the water-management system. Water managers have had public contacts and have been assessing receptivity of the clients to the proposed innovation, but this has not been their primary task. The managers have been concerned with cost factors, hazards, and the water supply/waste disposal.

The feedback can be supportive or it can signal opposition, and the conflict potential is one of the primary concerns about which managers desire more information. The feedback in Lubbock on the Canyon Lakes scheme was generally positive, and the water-management system was encouraged to continue in the direction it had chosen. In San Angelo, where this stage of the process was not actually engaged, some of the initial feedback was negative, and the system became even more cautious.

Stage III of the model presents a statement on water-management system strategy and a rationale for comparing wastewater reclamation and reuse to fluoridation. Absent, however, are statements on community conflicts; in the study sites, there were none, although perhaps San Angelo might have experienced conflict if the innovation had proceeded. The statements regarding strategies in community conflict are then basically projections by the actors as to what they would do if, for example, they were presenting direct potable re-use to the community. More definitive statements must await experience.

*Water-management system strategy (III-1)*. The innovation strategy of the water-management system arises from a number of factors and is "a distinctive combination of ends, means, and decision criteria."[53] It involves an interaction among the private and public goals of the components and assumes at least a minimum level of agreement. It includes considerations of environmental constraints of a biophysical and social nature, physical suitability, expected behavior of regulators, and anticipated public reactions.

Strategy also should involve a process of successive approximation, even if much of it is informal. The elements of the strategy should freely change as the system loops through cycles and obtains feedback. Chapter 3 contains detailed data for each site describing the elements of the strategies. This analysis aims at the water-management system's presentation of the re-use innovation to the public.

In Lubbock an announcement of the proposed innovation and a public information program touting local and regional economic and recreational

52. See John Romani, "The Administration of Public Health Services," in *Politics, Professionalism and the Environment*, ed. Lynton Caldwell, Environmental Studies, 3, (Bloomington, Indiana, Institute of Public Administration, 1967), pp. 1–21.

53. For an excellent study of water management in these terms, see White, *Strategies*.

54. See the strategies outlined in Charles Foster, *Waste Water Renovation for Water Supply: Its Political and Social Acceptability*, unpublished report, Baltimore, Maryland (January 1969).

benefits were the basic elements of issue presentation. Following initial groundwork (both technical and political), the Canyon Lakes Project received strong endorsement by the Council, and the same presentation was made to community groups. Proponents had high hopes for the Project and they accordingly launched a well-planned and dynamic effort. An important factor was that Lubbock had a model.

Given a model that was more experimental and involved some higher uses, Santee demanded a different approach to innovation; Santee chose a strategy of incrementalism which, because of suprasystem opposition, may well have been somewhat more incremental than the innovation proponents would have wished. Public reaction created no problems at Santee, but it may have been helped by this incremental strategy. Each incremental use triggered enthusiastic public response, which in turn bolstered the confidence of the water-management system. The managers recognized that this public approval strengthened their hand with the Department of Public Health.

In cases where the perceived hazard potential, and therefore possibilities of political conflict, are high, the water-management system may try an even more cautious strategy.[54] This could involve leaks and trial balloons (perhaps for incremental propositions) to probe public and other reaction. If trial balloons receive negative response, the system may wait and/or try to rearrange some elements of the strategy or environment (or both). If the response is positive, the system could then more openly propose a given level of use. Although this was not the case in any of the sites, it is a practical strategy in the hazard situations.

Primarily because of the analogy with the fluoridation experience, researchers in wastewater re-use generally cite the proposition that centralized, cohesive action by the water-management system increases the prospects for re-use adoption (the context implied is drinking but this is not always clear).

Whether the strategy assumptions are valid or the requirement for centralization activity real, the current prospects in most cities for administrative adoptions of wastewater re-use for very high-order uses, particularly drinking, are not good. The little evidence available from the study suggests that as the order of use increases, administrative willingness to adopt decreases. This holds for very high-order uses and for the pioneering stage of the innovation. Both technical and political actors in the water-management system would generally prefer to spread the risk through an extensive decision-making network, eventuating in a probable referendum. San Angelo probably would have held a referendum as much for reasons of assessing public preferences as for procuring the funds.

The expansion of the decision field would occur both because managers would want feedback on attitudes and because certain groups in the political environment would feel they had a stake in this decision. Even the stronger opponents for re-use would want wider sanctions than their own endorsements at the higher-use levels. The Public Health Department's approval, for example, was regarded by all as a sine qua non. This is also consistent, it should be noted, with the findings of Feldman in Chapter 6 on the situation in Israel.

The water managers emphasized public information programs and public approval, especially if drinking were to be involved. But unlike lower-order uses, including ornamental lakes, where client satisfaction could be achieved at low risk, the benefits of championing direct-potable re-use may not outweigh the costs. Moreover, it is obvious that at least a certain minimum level of agreement is necessary for the water-management system to act as a unit (and it must act as a unit even if not unanimously). This agreement would be harder to achieve as the use level rose to drinking.

Toward this point one should reemphasize the pivotal role of the public health department. Without this agency's approval, the water-management system would not continue to support the re-use innovation. And the evidence from Chapters 4 and 8 suggests that this support may be slow in coming. In the face of public health opposition, Chapter 7 makes clear that it would be practically impossible to convince the public.

Indicating what they "would do if they were going to propose wastewater reclamation for drinking," most managers emphasized extensive public information and risk-spreading even though such a course might seem to jeopardize prospects for adoption. They claimed that they had followed these guidelines even in lower levels of use and insisted that they would certainly follow them with higher-order uses.

Most if not all managers stressed need. As previously indicated, "need" is a flexible term; Colorado Springs and Lubbock emphasized need in their plans even though water adequacy was projected until the 1990's. With direct potable re-use, however, one would expect that the case for "need" would be clearer and more immediate. Even if the reclaimed water is to be added over an extended time period and through a system of buffers, the need argument must be impressive and the benefits clearly outlined. It seems that much of the general public was not convinced of the need for fluoridation, or saw the benefits as marginal and failed to support it when doubts were raised.

As in fluoridation, objections to wastewater reclamation for drinking are likely to be based on perceived threats to the quality goal, specifically the hazard potential. The water managers recognized this and therefore sought endorsements from the regulators of the quality goal. Although they mentioned scientists, suprasystem (state and federal government) technical personnel, and college professors, managers *emphasized* only the endorsements of medical personnel and relevant public health agencies. Some managers felt that the public would be more likely to trust a local doctor. Failure to receive public health endorsement would impair and probably halt consideration of re-use. Even without the legal problems of proceeding without public health approval, adverse public sentiment would pose a major obstacle. Managers regarded public sentiment as extremely important. They were cognizant of the conflict potential in negative public sentiment and had no desire to impair their public image, but they expressed interest in assessing these public preferences and using them as one guideline for formulating strategy. Probing and the environmental survey have already been outlined in Chapter 3. If in each site there were public information efforts, there were also efforts at feedback assessment.

In such efforts demonstration models would be central elements. The Santee model is very much tied into programs for reclaimed wastewater lakes at Antelope Valley, Tucson, and Lubbock. In each case the model was a source of support (among the technical and political components) in the early stages of the innovation and would aid in building public support when the public information programs began. The Santee model was for ornamental-recreation lakes; such a model for drinking would be of even more value. Chanute and Windhoek do not seem to be of much value as precedents. A model of an American city using reclaimed water for drinking would undoubtedly aid re-use prospects elsewhere considerably. Which will be the first cities? Tempting financial benefits (as in Santee research grants, facilities, and personnel subsidization) and image values (dynamic, innovative administration) are not without possibilities of conflict.

These obstacles are, however, not insurmountable. Some managers, in addition to assessing more general public preferences, seek to identify specific groups in the environment as positive or negative and to incorporate the stances of these groups in their considerations of strategy. Professional endorsements and support from various civic groups could serve to counteract the opposition from negative groups, although Chapter 7 suggests the conflict potential of even a small minority in the sensitive policy area of public health. In a referendum or situation involving wide participation, some managers identified the goal of isolating the virulent opposition and aiming the public information efforts at the sizable middle groups whose unified support guarantees desired results.

The basic premise of the public information effort is that it be factual, complete, and honest. The assumption behind isolating opponents is that their stance may be "strongly emotional, irrational, and based on causing fear and doubt". The behavior of certain groups in the fluoridation disputes helped to shape this assumption. The managers emphasized that they would want to be open and factual and outline the complete range of costs and benefits, and if there should be a hazard potential, to identify it. The major stress was on being meticulously well informed, and on reducing the chances of ever having to say, "I don't know."

These then are the major elements of a strategy for presenting the direct potable re-use (and other high-order uses) issue to the public. Administrative adoption or strong water-management system action in the face of an ambivalent or hostile environment seems unlikely. Regardless of the prospects of re-use adoption, risk-spreading and even referenda seem likely. The impact of perceived need on this whole process is speculative but would seem to improve the chances for re-use.

The hierarchy-of-use notion advocated in this book reduces these problems immensely. If replacement of the fresh supplies of major users with reclaimed water is possible, it may be the most suitable supply strategy. After all, it limits the hazard potential, protects both public and managerial perception of the quality goal, reduces the decision field, increases experience with reclamation processes, and minimizes the conflict potential. These are not trivial advantages.

One must emphasize, in closing, the conditional nature of the foregoing statements. They are most appropriate to the pioneering stage of an innovation. At Santee the strategies of implementation involved substantial caution. Subsequent experiences with similar situations may allow a reduction, but not an elimination, of this caution. Direct potable re-use may well constitute a similar process, but one must expect it to be slower and more problematic.

# Water Re-Use and Public Policy*

**10**

The present picture of water supply needs in major American cities balanced
against diminishing sources indicates an urgent need for carefully coordi-
natéd national policy to exploit fully the potential of water reclamation and
re-use. Policy statements by the American Water Works Association and the
Water Pollution Control Federation have stressed the importance of including water re-use in supply planning while at the same time underlining the
need for an ambitious research program aimed at resolving a number of
health hazards connected with the direct potable re-use of wastewater.
Nevertheless, there is the danger that other important considerations, particularly those which are behavioral, will be neglected over the short run as
the nation increasingly moves to new sources of urban water supply.

All policy recommendations rest upon an often unstated set of assumptions and guiding principles. These premises should be stated explicitly for
critical examination and debate. The studies of past innovations in water
re-use conducted by this project suggest a number of appropriate premises,
major findings, and, finally, modest policy recommendations.

*The following policy
statement reflects the
views of Stephen
Feldman, Roger E.
Kasperson, David
McCauley, and John T.
Reynolds.

## Premises

**1.** Traditional means of water-supply augmentation (for example, surface-
water exploitation, long-distance transport of water) and pollution abatement
(for example, point-source disposal of untreated or partially treated wastewater) are increasingly inadequate for urban water-management needs. There
must be a rapid development of watewater reclamation and re-use systems
in American metropolitan areas to meet water supply and wastewater disposal needs.

**2.** Recent federal legislation, particularly the Clean Water Amendments (PL
92-500) and the Safe Drinking Water Act, has substantially accelerated the
future pace of water re-use innovation. Even partial fulfillment of the former
will provide a high-quality effluent that will require only modest marginal
treatment costs to suffice for a number of re-use possibilities. At a time when
traditional means of water-supply augmentation are rapidly escalating in
price, this has enhanced the economic feasibility of wastewater reclamation
and re-use. In addition, the search for regional wastewater management

schema under Section 208 provides the groundwork for widespread planning innovation.

**3.** In managing their environments, societies should avoid, wherever possible, operation at the forefront of technologies which are poorly understood and which needlessly pose significant long-run risks to the health and well-being of their populations. The red flag in this case is the nuclear industry which has continuously pushed the development of technology into poorly understood and risky areas, thereby enlarging the total available risk, before consolidating the implementation of lower technology and its attendant risks. Such a rate of technological innovation needlessly enlarges uncertainty for only marginal economic gains. Where public water supplies are involved, such a process should be avoided.

**4.** The public health issues involved in the direct re-use of reclaimed waste-water for potable uses, whether planned or unplanned, continue to pose serious uncertainties related to contaminants (especially viruses and heavy metals), inadequate monitoring and fail-safe systems, deficiencies in backup water-quality laboratory analysis systems, and shortages in adequately trained operating personnel. The conversion of these uncertainties into definable risks whose general parameters are understood and the development of measurement instruments to test the parameters will not occur in the immediate or perhaps even in the short-term (10–15 years) future. There is also a pressing need for water-use standards that relate specifically to re-used water.

**5.** The highest quality of water should be reserved wherever possible for potable use, and the use of that water, conversely, should be minimized for lower-order uses whose quality needs are less.

The research on past water re-use adaptations and the attitudes and behavior of various actors uncovered in this study suggest a set of major findings relevant to policy recommendations.

## Major Findings

**1.** Economically efficient water re-use systems, particularly those which optimize capital investment over time to meet peaking needs by employing re-use as a standby or emergency system or which provide reclaimed water by direct piping to large-volume users, can be designed. Los Angeles County and Colorado Springs in the United States and the national re-use program in Israel provide convincing prototype models. Inadequate consideration is now given in the planning process to re-use possibilities partially because of the unavailability of methodology to compute the economic value of water re-use. A simulation model proposed herein at least partially meets that need.

**2.** Currently there is deep division in the managerial community over the desirability and feasibility of direct potable use over the short term and over land-disposal strategies of wastewater renovation and re-use for large cities. Although both consulting engineers and public health officials are conservative on the question of direct potable re-use, the latter are more resistant to the innovation and believe that much more time will be needed for routine adoption. Public health officials also tend to see public opposition and adverse political reaction as more formidable obstacles than do consulting engineers. Whereas public health officials are preoccupied with questions over the adequacy of technology and health issues, consulting engineers are primarily concerned with economic feasibility.

**3.** There is evidence that the conservatism of public health officials lies not only in their professional reservations over the health issues involved in direct potable re-use of wastewater but also in such behavioral bases as social origins, professional training and socialization, and occupational mobility which reinforce these evaluations. These attitudes will be slow to change, and there will be continued reluctance to engage in what are, in the professional milieu, high-risk activities.

**4.** Managers are, on the whole, highly informed about water re-use. Many have attended special conferences on the subject and visited the pioneering projects. Wastewater managers, rather than water-supply personnel, are likely to be the champions of water re-use. Resistance by water-supply managers is likely to be greatest where communication and coordination between water-supply and wastewater management are weak.

**5.** Water supply managers, like other bureaucrats, seek autonomy in their area of administrative discretion. Water reclamation and re-use may, because of the entrance of suprasystem officials and greater external regulation, create hostility among managers and foster a resistance to such innovation.

**6.** The impetus for innovations in water re-use will come chiefly from the technical-managerial community. Consistency in support and commitment among managers to such projects is essential for their success.

**7.** Education, past experience, and severity of need are all important determinants of public attitudes to the use of renovated wastewater. The major potential obstacle to public acceptance for potable re-use appears to lie less in psychological factors than in possible disagreement among the experts to whom the public looks for evaluation and guidance. Critical among these referents are public health officials and members of the medical profession.

**8.** If water reclamation and re-use systems utilize appropriate designs to ensure safety, employ sequential or staged experience with the hierarchy of uses, provide adequate information and preparation, and obtain a concert of support among experts, public acceptance is likely.

## Policy Recommendations

**1.** *A national comprehensive risk-assessment program*. A far-reaching and balanced risk-assessment effort is an urgent priority for the planned, direct potable use of reclaimed water. The American Water Works Association and the Water Pollution Control Federation have requested such a mass effort since 1971. The Safe Drinking Water Act and the present EPA analysis of contaminants in community water-supply systems will contribute important knowledge. But the fact of the matter is that current efforts are not commensurate with the magnitude of the problem given a near-term timetable for making water re-use systems available to American cities.

Even the efforts in progress tend to focus on certain types of risk assessment (viruses, heavy metals). Meanwhile, other problems—supportive fail-safe and monitoring technology, professional manpower shortages, the infrastructure of needed support facilities, problems of human error—apparently are receiving less searching analysis. The discrepancy between the increasing availability of reclaimed water as a result of PL 92-500 and the continuing lack of a comprehensive risk-assessment program to permit the re-use of this water is reason for serious national concern.

The present authors are cognizant of the argument that direct potable re-use systems can be designed which will in all probability provide a better product than that used in a number of extant water-supply systems, particularly those with inadvertent re-use. Such a view, however, seemingly accepts an existing evil as a norm for future water-supply systems. Wastewater re-use systems should emulate the best, not the worst, of urban water supplies. Needless health hazards can and should be avoided.

**2.** *Plural water supplies*. There must be a breakthrough in the sacrosanct status of a single grade of water for the wide discrepancy of urban uses. Even the planning efforts of the Army Corps of Engineers, revolutionary in some respects, have largely uncritically accepted this assumption. Despite widespread support of the concept among scientists and water professionals, there is currently a lack of effective leadership for innovation.

Wherever possible, the design of water supply systems should permit re-use for nonpotable needs (especially industry and agriculture). In all cities, direct piping to large-volume users should receive high priority in wastewater management efforts. Section 208 of PL 92-500, which calls for areawide wastewater management planning, provides an unparalleled opportunity for major innovations in this area. But unless federal leadership is forthcoming, it is unlikely that extensive multiple distribution systems will receive more than polite attention.

In rapidly growing cities it may prove economically feasible to institute dual distribution systems in new subdivisions or districts. Even in older cities, technological developments and density economies may make dual supply systems more economical than conventional wisdom suggests.

Utilization of a hierarchy of water supply would, by allocating reclaimed wastewater to the lower-order needs that constitute the bulk of demand, release the potable water supply for future expansion. At the same time, it would minimize the uncertainty involved in technological and managerial innovation while experience accumulates.

**3.** *Minimizing uncertainty through natural buffers*. There should be rapid development of technology in anticipation of eventual potable use of renovated wastewater. The present timetable for adoption of a direct potable re-use system suggests implementation by the end of the century in such places as Denver, Dallas, and perhaps Phoenix.

When these pioneering systems or other earlier emergency adoptions come on line, they should utilize wherever possible the indirect cycle in order to provide natural buffers against uncertainty. The closing of the loop should be designed in such a way as to exploit the imperfectly understood natural purification processes and dilution. The experience accumulated to date suggests that the natural buffers afforded by a return to indirect cycle provide a form of redundancy, an additional margin of safety, needed for early experience with new technological systems.

**4.** *A water re-use behavioral observatory*. There is a clear danger that technological development and health-related research will outstrip accumulated knowledge on the behavioral issues associated with water reclamation and re-use. Yet as the fluoridation experience and the present nuclear-energy program suggest, this is an area of prominent need if the nation is to exploit fully potential technological progress for future urban water-supply and wastewater management. Technology in water reclamation and re-use has greater capacity for change than managerial and public institutions and attitudes. Current behavioral research is fragmented among a number of governmental agencies, state water resources research centers, foundations, and other research programs. There is a pressing need for one center to monitor behavioral problems as they occur in the early stages of innovation and to conduct a coordinated basic research program directed to critical short-term and long-term priorities. The Observatory should also be responsible for an aggressive program of public education aimed at disseminating the results of its efforts to interested parties in the field of water reclamation and re-use.

**5.** *A national demonstration and dissemination program*. Section 208 of PL 92-500, which provides for areawide wastewater management planning, will, by broadening the alternative traditionally considered in wastewater management and emphasizing regional solutions, fill a much needed gap in water-planning. There is a need, however, for a new federal program specifically aimed at encouraging major innovations identified in 208 or other plans connected with water reclamation and re-use and disseminating their results. Without external support, the more imaginative and experimental of such alternatives may never be implemented. In particular support should be provided for:

(a) the development of varying scales and types of wastewater treatment plants with high degrees of reliability and new means of wastewater collection;

(b) alternatives in distribution systems, particularly those which allow re-
    claimed water to be used for lower-order needs, thereby releasing
    higher-quality water for potable uses;

(c) projects that produce multiple grades of water tailored to the needs of
    specific, large-volume urban users;

(d) coordination of land uses in regional planning designed to exploit re-use
    opportunities;

(e) integration of wastewater and water supply agencies in private consult-
    ing and public bodies;

(f) public education and information programs;

(g) efforts aimed at a change of attitude among managers;

(h) the development and application of new methodologies designed to
    evaluate the economic feasibility of water reclamation and re-use.

**6.** *Community decisions and water re-use.* Confronted by a possible adoption of water re-use, a community inevitably faces the question of how best to
proceed. To a considerable degree, the answer will vary with the individuality
of the particular community. The results of this study indicate that responsibility and success both suggest an open process that stresses a graduated
experience with reclaimed water up the hierarchy of uses, an ambitious program of public education and information, and consistent support among
experts and public officials. There is considerable potential for conflict in
high-order uses of reclaimed water. A fully informed and involved public
which is able to accumulate some experience with water re-use has the
greatest potential for community cooperation and approval. A sincere commitment to public education is also, it should be noted, the only responsible
course in a democratic society.

# Appendix 1
## Clark University Water Re-Use Study
## General Public Interview

Schedule Number
Site
Sex
   (m = 1, F = 2)
Statement
   (F = 1, d = 2)

1. How long have you lived in this city?

   1 = under 1 year
   2 = 1–5 years
   3 = 6–10 years
   4 = 11–20 years
   5 = 21–40 years
   6 = over 40 years

2. Could you please rate your water:

   _____ excellent to drink (1)     _____ a lot of cloudy material  (3)
   _____ average to drink  (2)      _____ some cloudy material      (2)
   _____ poor to drink     (3)      _____ clear                     (1)

   _____ average taste     (2)      _____ distinct color            (3)
   _____ good taste        (1)      _____ colorless                 (1)
   _____ poor taste        (3)      _____ slight color              (2)

   _____ no odor           (1)
   _____ strong odor       (3)
   _____ some odor         (2)

   9 = don't know

3. How often do you use bottled water in this household?

   1 = never
   2 = once or twice a year
   3 = once or twice a month
   4 = once or twice a week
   5 = once or twice a day
   9 = don't know

4. (If yes on no. 3) Is there any special reason for this?

   1 = special health problem
   2 = dissatisfaction with public supply
   3 = high mineral content desired
   4 = presence of fluoridation
   5 = for ironing
   6 = for mixing drinks
   7 = for infants, children
   8 = other
   9 = don't know

5. Have you ever heard of the process of treating municipal sewage so that it can be re-used?

   1 = yes
   2 = no
   9 = don't know

   If No to no. 5, go directly to Information Stem.

6. (If yes to no. 5) Have you heard of reclaimed water being used for:

   Industry?              _____
7. Agriculture?          _____ 1 = yes
8. Fishing?              _____ 2 = no
9. Swimming?            _____ 3 = don't know
10. Drinking?           _____
                        Code for total number of
                        uses identified

11. How did you feel about the idea when you first heard about it? (refer to highest order use)

    1 = Disapprove strongly
    2 = Disapprove
    3 = Neutral
    4 = Approve
    5 = Approve strongly
    9 = don't know

12. How do you feel about it now?

    1 = Disapprove strongly
    2 = Disapprove
    3 = Neutral
    4 = Approve
    5 = Approve strongly
    9 = don't know

Code Relative Change:

1 = −4
2 = −3
3 = −2
4 = −1
5 = no change
6 = +1
7 = +2
8 = +3
9 = +4

13. (If changed), what made you change your mind?
    (SCALE TO BE DEVELOPED IN FIELD)

14. With whom have you talked about it?

    1 = family or relatives
    2 = friends
    3 = work or school associates
    4 = elected officials
    5 = medical personnel
    6 = public health officials
    7 = water or sanitary managers
    8 = other
    9 = nobody
    ADMINISTER INFORMATION STEM.

INFORMATION STEM F

Reclaimed water is municipal sewage which has been cleaned and treated. Such water would be used to add to the present water supply. Reclaimed water meets the standards for drinking water of the U.S. Public Health Service and can be produced at reasonable cost. It has been used for fishing and swimming in Santee, California, and for drinking in the industrial town of Windhoek, South West Africa. Because it improves the quality of the wastewater returned to rivers and lakes, water reclamation also helps to reduce water pollution.

15. Your community is going to have to add to its water supply to keep up with demand. This can be done by either expanding your present sources or by adding renovated municipal wastewater to your present system. Renovated water would be cheaper; would you prefer to use it or to pay the extra money to stay with your present sources?

    1 = Use renovated water
    2 = Pay extra for present sources
    9 = don't know

16. (If pay extra) Would you be willing to pay (25%, 50%, 100%) more on your water bill to keep your present sources of water? (Circle figure given)

    1 = yes
    2 = no
    3 = inappropriate
    9 = don't know

17. What is the occupation of the head of the house?

    _____

18. What is your religion?

    1 = Protestant
    2 = Roman Catholic
    3 = Jewish
    4 = Orthodox
    5 = Agnostic–Atheist
    6 = Other
    9 = don't know

19. Would you please rank the following uses of reclaimed water on this scale?

| | Favorable | | Neutral | Unfavorable | |
|---|---|---|---|---|---|
| | +2 | +1 | 0 | −1 | −2 |
| Swimming | ___ | ___ | ___ | ___ | ___ |
| Industrial Cooling | ___ | ___ | ___ | ___ | ___ |
| Drinking Water | ___ | ___ | ___ | ___ | ___ |
| Washing Clothes | ___ | ___ | ___ | ___ | ___ |
| Cooking | ___ | ___ | ___ | ___ | ___ |
| Golf Course Irrigation | ___ | ___ | ___ | ___ | ___ |
| Vegetable Irrigation | ___ | ___ | ___ | ___ | ___ |
| Fishing | ___ | ___ | ___ | ___ | ___ |

For Coding (−2 = 1, −1 = 2, 0 = 3, +1 = 4, +2 = 5)

20. Would you please rank the following possible ways of incorporating reclaimed water into your present water supply? Rank your preferred way with a 1, and your second choice with a 2.

    Aquifer Recharge (see diagram)
    Mixing reclaimed water with water in a reservoir (see diagram)

21. The opinions of the following people might be involved in a decision to accept the use of reclaimed water. Please rank them, listing first (1) the one to whose approval you would give the most importance, listing second (2) the next most important and so on.

Local Doctors
Mayor/City Manager
U.S. Public Health Officials
Editorials (TV/Newspaper)
Engineers, Scientists
Water Supply Managers

## Opinion and Attitude Survey

*Instructions*: The questions below consist mostly of the beginnings of sentences which you are asked to complete. Simply finish each sentence with whatever comes first into your head. If a way to complete a sentence doesn't come to mind right away, skip that sentence and go on to the next, and come back to it later. If you have any questions, please ask the interviewer for clarification.

1. If I were eating an apple and dropped it on the floor, I would
   _____

2. I believe that luck _____
   _____

3. Of the following two statements, which is closer to your own attitude toward your health? (Please check one.)
   I tend to take my health for granted and rarely give it thought until
   _____ something goes wrong and I don't feel good.
   I pay considerable attention to my health and am careful to take good
   _____ care of myself.

4. The solution of pollution problems lies in _____
   _____

5. I believe that the practice of seeding clouds with silver nitrate in order to make it rain _____
   _____

6. Given a choice between the whipping cream that comes in an aerosol can and that cream which you have to whip yourself, I would buy
   _____ because _____
   _____

7. In making progress, man has _____
   _____

8. I believe that fluoridation of water _____
   _____

9. Getting ahead in the world results from _____
   _____

10. I believe that vaccination _____
    _____

*Check education, age, and income categories most appropriate to you.*

11. What is the highest grade in school which you have completed?
    _____ 1 = None
    _____ 2 = Some grade school or grade school
              graduate
    _____ 3 = Some high school or high school graduate
    _____ 4 = Trade School
    _____ 5 = Some college or college graduate (4 years)
    _____ 6 = Some graduate work or graduate or professional degree
    _____ 7 = Other
    _____ 8 = Don't know.

12. Could you indicate your total family income (approximately) for 1969?
    _____ 1 = Under $5,000
    _____ 2 = $5,001–$10,000
    _____ 3 = $10,001–$15,000
    _____ 4 = $15,001–$20,000
    _____ 5 = more than $20,000
    _____ 9 = don't know

13. What is your approximate age?
    _____ 1 = Under 21          _____ 4 = 51–65
    _____ 2 = 21–35             _____ 5 = over 65
    _____ 3 = 36–50

# Appendix 2
# A Note on the Thematic Apperception Test
John Sims

The projective method used to gather data analyzed in Chapter 8 is a variant of the Thematic Apperception Test, or TAT.[1] For readers unfamiliar with the TAT and its psychological rationale, these are briefly discussed and examples of test data are presented.

1. H. A. Murray, *Thematic Apperception Test: Pictures and Manual* (Cambridge, Massachusetts, Harvard University Press, 1943).

The standard test consists of a series of ambiguous pictures about which the subject is asked to tell stories. Since the stimuli of the pictures are ambiguous, the thoughts, feelings, attitudes, and actions with which the storyteller endows the persons portrayed in the pictures are indicative of the psychology of the storyteller himself. What happens, in effect, when asked to interpret an ambiguous stimulus, is that one projects onto the stimulus one's view of himself and one's view of the world.

The essential logic for using a projective technique as opposed to direct methods such as explicit questioning is to prevent the subject from mobilizing his defenses, both conscious and unconscious, when responding. Thus, for example, if you ask an individual how he gets along with his spouse (or with his parents or children), the social and psychological pressures to answer positively are considerable. But if you show him a picture of a husband and wife (or of a parent and child) and ask him to tell a story about it, it is easy for him to "tell the truth" about his own relationship because the person's pictures are "not they."

Experience has shown that particular pictures tend to produce stories that reveal particular psychological dimensions—for example, achievement motivation, or attitudes toward authority. This allows the researcher to select pictures from the standard TAT set, or to design new pictures, which elicit data specifically relevant to his interests.

The picture used for these studies portrays a group of seven adult males in business dress grouped in various attitudes around a conference table. With the presentation of the picture, subjects were read the following instructions:

> This is a picture of a meeting in a mayor's office which he has called to discuss the possibility of coping with an impending local water shortage through the use of renovated wastewater. I would like you to use your imagination and tell me a story about it. Who do you think are the various persons attending the meeting? What is going on at the moment?

What are the men thinking and feeling and saying? How do you think this situation will turn out?

Clearly, the instructions reduce the ambiguity of the stimulus; indeed, what is initially going on in the picture is explicitly defined. The stimulus (now a combination of picture and instructions) *remains*, however, ambiguous in regard to the areas of investigator interest—that is, whom will the subject identify as attending the meeting, what will their attitudes be toward the use of renovated wastewater, and how will the situation be resolved? Thus *these* questions will be answered by the projections of the storyteller.

The two stories presented below illustrate important similarities and differences in how subjects perceive a situation in which the use of renovated wastewater is being seriously considered.

**Subject 24:** *This is the mayor. This one is the water commissioner and next to him is his assistant, who is a consulting engineer. This is the public health official of the city. The next is the fire chief; there's not much he can do about the situation. The other two are aldermen. It looks like they are fully aware of the problem at this point. The engineer is explaining the various ways to overcome the problem. The water commissioner is satisfied that the engineer is explaining the situation and the solution properly. The public health official is also understanding of the problem. There is not too much he can do at this point as far as the actual water supply goes. He's advised them of the possible health effects and the safeguards that must be taken if the water is to be re-used. The fire chief has been called upon as a courtesy, not much he can do. The mayor has been listening to the advice of the people he has called in, and he will rely on the expertise of these people to recommend the best solution. He is worrying about the repercussions as far as having to tell the people that they are re-using water. The aldermen appear concerned and worried about the situation, but they are lay people and know practically nothing about the problem or its solution. The water department will present its recommendation and the health department will accept with the provision that safeguards be placed to see that the proposal is carried out in a safe manner. They will accept the proposal.*

**Subject 56:** *This man is from the state health department. He is concerned with the safeguards and the public acceptance. Has a show-me attitude. Next man is the consulting engineer. He's proposed the idea and is trying to sell it as a viable alternative. The mayor is concerned with the political implications and the cost. The water treatment plant operator is disgusted with the proposal because of the expense and personal repulsion. This fellow is the head of the city department of works–he's concerned with how to provide controls and safeguards as well as the cost. This man is the city health official who is concerned with the health aspects and the public acceptance. And finally, this is the president of the city council and represents the citizens' groups. He is opposed to it. They will go back to try to find another solution and will not stick their necks out at this time. They will see if some other community does it first.*

# Bibliography

Ackerman, Neil. "Water Reuse in the United States." Unpublished M.A. thesis, Department of Geography, Southern Illinois University (1971).

American Water Works Association. "On the Use of Reclaimed Wastewater as a Public Water Supply Source, AWWA Policy Statement," *Journal of the American Water Works Association*, 63 (1971), 609.

American Water Works Association and Water Pollution Control Federation. "Joint AWWA-WPCF Statement," *Journal of the American Water Works Association*, 65 (1973), 700.

American Society of Civil Engineers, Sanitary Engineering Division, Committee on Environmental Quality Management. "Engineering Evaluation of Virus Hazard in Water," *Journal of Sanitary Engineering*, ASCE, SA1, 96 (1970) 111–161.

Amramy, A. "Reuse of Municipal Wastewater," in *Proceedings of the International Conference on Water for Peace*, 2 (1967), 423–424.

*Antelope Valley Press*. August 10, 1962; August 30, 1962; February 24, 1963; December 2, 1965.

Athanasiou, Robert B., and Hanke, Steve H. "Social Psychological Factors Related to the Adoption of Reused Wastewater as a Potable Water Supply," in *Urban Demands for Natural Resources*, ed. Western Resources Conference (Denver, University of Denver, 1970), pp. 113–125.

Bachrach, Peter, and Baratz, Morton S. *Power and Poverty: Theory and Practice* (London, Oxford University Press, 1970).

Bain, Joe S., Caves, Richard E., and Margolis, Julius. *Northern California's Water Industry* (Baltimore, Johns Hopkins Press for Resources for the Future, 1966).

Barnett, Homer G. *Innovation, the Basis of Cultural Change* (New York, McGraw Hill, 1953).

Baumann, Duane D. "Perception and Public Policy in the Recreational Use of Domestic Water Supply Reservoirs," *Water Resources Research*, 5 (1969), 543–555.

———. *The Recreational Use of Domestic Water Supply Reservoirs: Perception and Choice*, University of Chicago Department of Geography Research Papers, 121 (Chicago, 1969).

———, and Dworkin, Daniel. *Planning for Water Reuse* (Indianapolis, Holcomb Research Institute, Butler University, 1975).

———, and Kasperson, Roger E. "Public Acceptance of Renovated Wastewater: Myth and Reality," *Water Resources Research*, 10 (1974), 667–674.

————, and Sims, John. "Renovated Wastewater for Drinking: The Question of Public Acceptance," *Water Resources Research*, 10 (1974), 675–682.

Baummer, J. C., and Robertson, P. G., eds. *Reclamation of Wastewater for Municipal Water Supply* (Baltimore, Department of Geography, Johns Hopkins University, 1973).

Beard, L. R. *Method for Determination of Safe Yield and Compensation Water from Storage Reservoirs*, U.S. Army Corps of Engineers Hydrologic Engineering Center Technical Papers, 3. (Sacramento, The Center, 1965).

Becker, Howard S.; Greer, Blanche; Hughes, Everett C.; and Strauss, A. *Boys in White* (Chicago, University of Chicago Press, 1961).

Bell, Daniel. *The Coming of Post-Industrial Society: A Venture in Social Forecasting* (New York, Basic Books, 1973).

Bell, Gerald, ed. *Transmission of Viruses by the Water Route* (New York, John Wiley and Sons, 1965).

Berthouex, P. M., and Polkowski, L. B. "Cost-Effectiveness Analysis of Wastewater Reuses," *Journal of Sanitary Engineering*, ASCE, SA6, 98, (1972), 869–881.

Bishop, A. Bruce, et al. *Evaluating Water Reuse Alternatives in Water Resources Planning* (Springfield, Va., National Technical Information Service, 1974).

Black and Veatch (Consulting Engineers). *Report on Cheyenne Canyon Booster District and Templeton Service Area for Colorado Springs, Colorado* (Kansas City, Mo., 1972).

————. *Report on Water Distribution System for Colorado Springs, Colorado, 1969 Revision* (Kansas City, Mo., 1969).

Blake, Nelson M. *Water for the Cities* (Syracuse, Syracuse University Press, 1956).

Bonderson, Paul. "Quality Aspects of Wastewater Reclamation," *Journal of Sanitary Engineering*, ASCE, SA5, 90 (1964), 1–8.

Bonne, J., and Grinwald, Z. "The Estimated Growth of Water Consumption in Israel," in *Water in Israel*, Part A, ed. Israel Ministry of Agriculture (Tel Aviv, 1973).

*A Brief History of the Water Supply System for Lubbock, Texas*. [n.d.]

Bruvold, William H. "Affective Response toward Uses of Reclaimed Water," *Experimental Publication System*, Issue 3 (December, 1969), Manuscript 1170, pp. 1–12.

————. "Public Opinion and Knowledge Concerning New Water Sources in California," *Water Resources Research*, 8 (1972), 1145–50.

————, and Ongreth, Henry J. "Public Use and Evaluation of Reclaimed Water," *Journal of the American Water Works Association*, 66 (1974), 294–297.

————, and Ward, Paul C. "Public Attitudes toward Uses of Reclaimed Wastewater," *Water and Sewage Works*, 117 (1970), 120–122.

————. "Using Reclaimed Wastewater—Public Opinion," *Journal of the Water Pollution Control Federation*, 94 (1972), 1690–96.

Buswell, A. M., et al. *The Depth of Sewage Filters and the Degree of Purification*, Illinois Division of State Water Survey Bulletins, 26 (Springfield, Ill., 1928).

Cannon, Daniel. "Industrial Reuse of Water: An Opportunity for the West," in *Water*, ed. Roma McNickel, Western Resources Conference, 1963 (Boulder, University of Colorado Press, 1964), pp. 69–78.

Cecil, Lawrence K., ed. *Complete Water Reuse: Industry's Opportunity*, Proceedings of the National Conference on Complete Water Reuse, American Institute of Chemical Engineering, April 1973 (New York AICE, 1973).

Clark, Peter and Wilson, James. "Incentive Systems: A Theory of Organizations," *Administrative Science Quarterly*, 6 (1961), 129–166.

Clawson, Marion. *Suburban Land Conversion in the United States* (Baltimore, Johns Hopkins Press, 1973).

Cohen, Michael; March, James; and Olsen, Johan. "A Garbage Can Model of Organizational Choice," *Administrative Science Quarterly*, 17 (1972), 1–25.

Colorado Springs, Colorado, Department of Public Utilities (Sewer Division). "City of Colorado Springs Sewage Treatment System." [n.d.]

Costantini, Edmund, and Hanf, Kenneth. *Environmental Concern and Political Elites: A Study of Perception, Background, and Attitudes*. University of California at Davis, Institute of Governmental Affairs, Research Reports, 21 (Davis, California, 1971).

Council on Environmental Quality. *Environmental Quality—The Fifth Annual Report of the Council on Environmental Quality* (Washington, D.C., 1974).

Craik, Kenneth. "The Environmental Dispositions of Environmental Decision-Makers," *Annals* of the American Academy of Political and Social Sciences, 389 (1970), 91–97.

Crain, Robert L.; Katz, Elihu; and Rosenthal, Donald B. *The Politics of Community Conflict: The Fluoridation Decision* (Indianapolis, Bobbs-Merrill, 1969).

Crenson, Matthew. *The Un-Politics of Air Pollution* (Baltimore, Johns Hopkins Press, 1971).

Culp, Gordon L.; Culp, Russell L.; and Hamann, Carl L. "Water Resource Preservation by Planned Recycling of Treated Wastewater," *Journal of the American Water Works Association*, 65 (1973), 641–647.

Culp, Russell L. "Uses of Reclaimed Water," *Water and Wastes Engineering*, 6 (1969), 54.

————. "Wastewater Reclamation at South Tahoe Public Utilities District," *Journal of the American Water Works Association*, 60 (1968), 84–94.

————, and Culp, Gordon. *Advanced Wastewater Treatment* (New York, Van Nostrand, 1971).

————, and Moyer, Harlan. "Wastewater Reclamation and Export at South Tahoe," *Civil Engineering*, 41 (1969), 38.

Darr, Peretz [P. Dalinsky]. "Intersectoral Competition for Brackish Waters" (Tel Aviv, 1972). (unpublished Tahal memorandum)
———. "Socio-Ecologic Aspects of Closed Wastewater Utilization Systems," *Proceedings of the Fifth Scientific Conference of the Israel Ecologic Society* (Tel Aviv, 1974).
———; Feldman, Stephen; and Kamen, C. *Operational Approaches to Urban Water Supply: A Religious Analysis of Water Supply Capacity in Israel* (Rotterdam, Holland, Rotterdam University Press, 1976).
Davis, George, and Wook, Leonard. *Water Demands for Expanding Energy Developments*, U.S. Geological Survey Circulars, 703 (Reston, Va., U.S. Geological Survey, 1974).
Dworkin, Daniel, and Baumann, D. *An Evaluation of Water Reuse for Municipal Supply* (Alexandria, Va., Institute for Water Resources, U.S. Army Corps of Engineers, 1974).
Dworsky, Leonard B. *The Nation and Its Water Resources* (Washington, D.C., National Technical Information Service, 1962).
Edelman, Murray. *The Symbolic Uses of Politics* (Urbana, University of Illinois Press, 1964).
Egelund, Duane R. "Land Disposal I: A Giant Step Backward," *Journal of the Water Pollution Control Federation*, 45 (1973), 1465–75.
Eichhorn, Robert. "The Student Engineer," in *The Engineer and the Social System*, ed. Robert Perrucci and Joel Gerstl (New York, Wiley, 1969).
*Environment* Staff. "The Wastewater Tide Ebbs Slowly," *Environment*, 15 (1973), 34.
Fair, G. W.; Geyer, J. C.; and Okun, D. A. *Water Supply and Wastewater Removal* (2 vols., New York, Wiley, 1966).
Feinmesser, A. *Survey of Sewage Collection Treatment and Utilization* (Tel Aviv, Israel Ministry of Agriculture, Water Commission, 1971).
Feldman, Stephen. "Artificial Rainfall Stimulation: A Comparative Analysis of Israel and the U.S. Southwest." (unpublished paper)
Firey, Walter. *Man, Mind and Land* (Glencoe, Ill., Free Press, 1960).
Flack, Ernest. "Urban Water: Multiple Use Concepts," *Journal of the American Water Works Association*, 63 (1971) 644–646.
Fleming, Robert, and Jobes, Harold. "Water Reuse: A Texas Necessity," *Journal of the Water Pollution Control Federation*, 41 (1969), 1564–69.
Foster, Charles. "Waste Water Renovation for Water Supply: Its Political and Social Acceptability." Unpublished report prepared for the Advanced Water Sciences and Management Project, Johns Hopkins University (Baltimore, Md., January 1969).
Frederickson, H. George, and Magnas, Howard. "Comparing Attitudes toward Water Pollution in Syracuse," *Water Resources Research*, 4 (1968), 877–889.
Freese, Nichols, and Endress (Consulting Engineers). *Feasibility Report on the Canyon Lakes Project* (Fort Worth, Texas, 1969).
Fuhrman, Ralph E. "Adaption of Known Principles and Techniques of Waste Water Management to Specific Environmental Situations and Geo-

graphical Conditions," *Journal of the Water Pollution Control Federation*, 68 (1969), 619–621.

Galbraith, John K. *Economics and Public Purpose* (Boston, Houghton Mifflin, 1973).

Gallup Poll. "Water Quality and Public Opinion," *Journal of the American Water Works Association*," 65 (1973), 513–519.

Gamson, William. "The Fluoridation Dialogue: Is It Ideological Politics?," *Public Opinion Quarterly*, 26 (1962), 526–537.

Garrison, Charles, and Hill, Duane. "The Dynamics of Public Roles in the Selection of Revenue Sources in Local Water Administration," *Water Resources Research*, 3 (1967), 949–962.

Gavis, Jerome. *Wastewater Reuse* (Springfield, Va., National Technical Information Service, 1971).

Gerstl, Joel E., and Hutton, S. P. *Engineers: The Anatomy of a Profession* (London, Tavistock, 1966).

Goldman, Charles R; McEvoy, James III; and Richerson, Peter J. *Environmental Quality and Water Development* (San Francisco, W. H. Freeman, 1973).

Gore, William. *Administrative Decision Making* (New York, Wiley, 1964).

Graeser, Henry J. "Dallas Wastewater Reclamation Studies," *Journal of the American Water Works Association*, 63 (1971), 634–640.

———. "Water Reuse: Resource of the Future," *Journal of the American Water Works Association*, 66 (1974), 575–578.

Griliches, H. Zvi. "Research Expenditure, Education and the Aggregate Agricultural Production Function," *American Economic Review*, 54 (1964), 961–974.

Hallock, Robert, and Ziebell, Charles. "Feasibility of Sport Fishery in Tertiary Treated Wastewater," *Journal of the Water Pollution Control Federation*, 42 (1970), 1656–65.

Haney, Paul O., and Hamann, Carl L. "Dual Water Systems," *Journal of the American Water Works Association*, 57 (1965), 1073–99.

Hanke, Steve H. "Demand for Water under Dynamic Conditions," *Water Resources Research*, 6 (1970), 1253–61.

Hansen, C. A. "Standards for Drinking Water and Direct Reuse," *Water and Wastes Engineering*, 6 (1969), 44–45.

Hart, Henry. "Crisis, Community, and Consent in Water Politics," *Law and Contemporary Problems*, 22 (1957), 510–537.

Hartman, L. M., and Seastone, D. A. "Alternative Institutions for Water Transfers: The Experience in Colorado and New Mexico," *Land Economics*, 34 (1963), 31–43.

Haynes, M. Alfred. "Professionals and the Community," *American Journal of Public Health*, 60 (1970), 519–523.

Headstream, Marcia; Wells, Dan M.; and Sneazy, Robert M. "The Canyon Lakes Project," *Journal of the American Water Works Association*, 67 (1975), 125–127.

Heaton, R. D., et al. "Progress toward Successive Water Use in Denver." Paper presented at the Water Pollution Control Federation Annual Meeting (Denver, 1974). (mimeographed)

Hirshleifer, Jack; DeHaven, James C.; and Milliman, Jerome W. *Water Supply: Economics, Technology and Policy* (rev. ed., Chicago, University of Chicago Press, 1969).

Houser, E. W. "Santee Project Continues to Show the Way," *Water and Wastes Engineering*, 7 (1970), 40–44.

Howards, Irving, and Kaynor, Edward. *Institutional Patterns in Evolving Regional Programs for Water Resource Management* (Amherst, University of Massachusetts, 1971).

Howe, Charles W. "Municipal Water Demand," in *Forecasting the Demands for Water*, ed. W. R. Derrick Sewell, et al. (Ottawa, Canada, Department of Energy, Mines and Resources, 1968).

————, and Linaweaver, F. P., Jr. "The Impact of Price on Residential Water Demand and Its Relation to System Design and Price Structure," *Water Resources Research*, 3 (1967), 13–32.

Hutchins, W. A. *Municipal Waste Facilities in the United States* (Washington, D.C., Environmental Protection Agency, 1972).

Inkeles, Alex. "Social Structure and Socialization," in *Handbook of Socialization and Research*, ed. D. A. Goslin (Chicago, Rand McNally, 1969), pp. 615–632.

International Union of Pure and Applied Chemistry. *Re-Use of Water in Industry* (London, Butterworths, 1963).

Israel, Ministry of Agriculture, ed. *Water in Israel*, Part A (Tel Aviv, 1973).

Jarrett, Henry, ed. *Environmental Quality in a Growing Economy* (Baltimore, Johns Hopkins Press, 1966).

Johnson, James F. *Renovated Waste Water: An Alternative Source of Municipal Water Supply in the United States*, University of Chicago Department of Geography Research Papers, 135 (Chicago, 1971).

Kally, E. "Cost of Conveying Water in Pressure Mains" (Tel Aviv, 1972). (unpublished Tahal memorandum)

————. "Israel's Water Economy and Its Problems in the Early Seventies," in *Water in Israel*, Part A, ed. Israel Ministry of Agriculture (Tel Aviv, 1973).

Kardos, Louis T.; Sopper, William E.; and Myers, Earl A. "A Living Filter for Sewage," in *Yearbook of Agriculture, 1968*, ed. U.S. Department of Agriculture (Washington, D.C., Department of Agriculture, 1969), pp. 197–201.

Kasperson, Roger E. "Environmental Stress and the Municipal Political System," in *The Structure of Political Geography*, ed. Roger E. Kasperson and Julian V. Minghi (Chicago, Aldine, 1969).

————. "Political Behavior and the Decision-Making Process in the Allocation of Water Resources between Recreational and Municipal Uses," *Natural Resources Journal*, 9 (1969), 176–211.

Kearns, John T. "Water Conservation and Its Application to New England," *Journal of The American Water Works Association*, 58 (1966), 1379–1384.

Kneese, Allen V., and Smith, Stephen C., eds. *Water Research* (Baltimore, Johns Hopkins Press, 1966).

Knutson, A. L. *The Individual, Society, and Health Behavior* (New York, Russell Sage, 1965).

Koenig, Louis. *Studies Relating to Market Projections for Advanced Waste Treatment*, U.S. Department of the Interior Publication, WP-20-AWTR-17 (Washington, D.C., 1966).

Lambie, John. *Waste Water Reclamation Project for Antelope Valley Area: Final Report* (Los Angeles, County Engineering Office, 1968).

Leif, Harold, and Fox, Renée C., eds. *The Psychological Basis for Medical Practice* (New York, Harper and Row, 1963).

Levinson, Daniel J. "Medical Education and the Theory of Adult Socialization," *Journal of Health and Social Behavior*, 7 (1967), 253–265.

Levite, G. E., ed. *Developments in Desalination Technology in Israel* (Jerusalem, National Council of Research and Development, 1971).

Linaweaver, F. Pierce, Jr.; Geyer, John C.; and Wolff, Jerome B. *A Study in Residential Water Uses* (Baltimore, Department of Environmental Engineering Science, Johns Hopkins University, 1966).

Linstedt, K. D.; Miller, K. J.; and Bennett, E. R. "Metropolitan Successive Uses of Available Water," *Journal of the American Water Works Association*, 61 (1969), 610–615.

Liu, O. C. *Effect of Chlorination on Human Enteric Virus in Partially Treated Water from Potomac Estuary*, Northeastern Water Hygiene Laboratory Progress Reports, November 1971 (Washington, D.C., E.P.A., 1971).

Long, William N., and Bell, Frank A., Jr. "Health Factors and Reused Waters," *Journal of the American Water Works Association*, 64 (1972), 220–225.

*Los Angeles Times*. June 21, 1967.

Loveridge, Ronald. *City Managers in Legislative Politics* (Indianapolis, Bobbs-Merrill, 1971).

Lubbock, Texas, Planning Department. *Canyon Lakes Project: Goal, Program, Policies* (Lubbock, n.d.).

———. *An Expanding Lubbock . . .: Reclaimed Water for a Growing City* (Lubbock, 1968).

*Lubbock-Avalanche Journal*. February 13, 1968; April 25, 1968; May 13, 1968; May 11, 1970; April 29, 1971.

Ludwige, F. H.; Kazmierczak, E.; and Carter, R. "Waste Disposal and the Future of Lake Tahoe," *Journal of the Sanitary Engineering Division*, ASCE, SA3, 90 (1964), 27–51.

McCabe, L. J. "Health Aspects of Reusing Wastewater for Potable Purposes." Paper presented at American Water Works Association meetings (June 1975). (mimeographed)

McDermott, James H. "Virus Problems in Water Supplies," *Water and Sewage Works*, 122 (May and June 1975), 71–73, 76–78.

McIver, Ian. "Municipal Water Supply in Grand River Basin Region," in *Perceptions and Attitudes in Resources Management*, ed. W. R. Derrick Sewell and Ian Burton (Ottawa, Canada, Policy Research and Coordination Branch, Department of Energy, Mines, and Resources, 1971).

McKean, Roland. "Costs and Benefits from Different Viewpoints," in *Public Expenditures in the Urban Community*, ed. Howard Schaller (Baltimore, Johns Hopkins Press for Resources for the Future, 1962).

McNickel, Roma, ed. *Water*, Western Resources Conference, 1963 (Boulder, University of Colorado Press, 1964).

Marine, G. "California Water Plan: The Most Expensive Faucet in the World," *Ramparts*, 8 (May 1970), 34–41.

Marshall, Hubert. "Politics and Efficiency in Water Development," in *Water Research*, ed. Allen V. Kneese and Stephen C. Smith (Baltimore, Johns Hopkins Press, 1966), pp. 291–310.

Marx, Jean. "Drinking Water: Another Source of Carcinogens?," *Science*, 186 (November 20, 1974), 809–811.

Marx, Wesley. *Man and His Environment: Waste* (New York, Harper and Row, 1971).

Mausner, Judith A., and Bahn, Anita K. *Epidemiology* (Philadelphia, W. B. Saunders, 1974).

Medbery, H. Christopher. "Preparing for Tomorrow's Needs," *Journal of the American Water Works Association*, 58 (1966), 930–938.

Merrell, John C., Jr., et al. *The Santee Recreation Project. Santee, California. Final Report*, U.S. Federal Water Pollution Control Federation Publication WP-20-7 (Washington, D.C., 1967).

————, and Katko, Albert. "Reclaimed Wastewater for Santee Recreational Lakes," *Journal of the Water Pollution Control* Federation, 38 (1966), 1310–18.

————; Katko, Albert; and Pintler, H. E. *The Santee Recreation Project. Santee, California, Summary Report 1962–1964*, U.S. Public Health Service, Publication 999-WP-27 (Washington, D.C., USPHS, 1965).

Metzler, Dwight F., et al. "Emergency Use of Reclaimed Water for Potable Supply at Chanute, Kansas," *Journal of the American Water Works Association*, 50 (1958), 1021–57.

————, and Russelmann, Heinz B. "Wastewater Reclamation as a Water Resource," *Journal of the American Water Works Association*, 60 (1968), 95–102.

Middleton, Frank M. "Municipal Drinking Water and Wastewater—A National Problem." Paper presented at a seminar on Water and Sewage Problems in Western New York (Rochester, New York, May 23, 1963).

Milgram, Stanley. "The Experience of Living in Cities," *Science*, 167 (March 13, 1970), 1461–68.

Murray, H. A. *Thematic Apperception Test: Pictures and Manual* (Cambridge, Mass., Harvard University Press, 1943).

Muskegon, Michigan, Muskegon County Board and Department of Public Works. *Engineering Feasibility Demonstration Study for Muskegon County Wastewater Treatment Irrigation System*, Water Pollution Control Series 11010 FMY (Washington, D.C., U.S. Department of the Interior, Federal Water Quality Administration, 1970).

National Academy of Sciences. *Water and Choice in the Colorado Basin*, NAS Publications, 1689 (Washington, D.C., National Academy of Sciences, 1968).

National Technical Advisory Committee on Water Quality Criteria. *Report of the Committee on Water Quality Criteria* (Washington, D.C., Government Printing Office, 1972).

*New York Times*. December 18, 1949.

Ockershavsen, R. W. "In-Plant Usage Works and Works," *Environmental Science and Technology*, 8, No. 5 (1974), 420–423.

Okun, Daniel A. "Alternatives in Water Supply," *Journal of the American Water Works Association*, 61 (1969), 215–224.

———. "Planning for Water Reuse," *Journal of the American Water Works Association*, 65 (1973), 617–622.

———, and McTunkin, F. "Feasibility of Dual Water Supply Systems." Paper presented to the 7th Annual American Water Works Association Meeting (October, 1971).

Owen, Langdon W. "Ground Water Management and Reclaimed Water," *Journal of the American Water Works Association*, 60 (1968), 135–144.

Pagorski, Albin D. "Is the Public Ready for Recycled Water?," *Water and Sewage Works*, 121 (1974) 108–109.

Parizek, R. R., et al. *Waste Water Renovation and Conservation*, Penn. State Studies, 23 (University Park, Pennsylvania State University, 1967).

Parkhurst, John D. "A Plan for Water Re-Use." Report prepared for the directors of the County Sanitation Districts of Los Angeles County, July 1963 (Los Angeles, 1963). (mimeographed)

———. "Wastewater Reuse—A Supplemental Supply," *Journal of Sanitary Engineering*, ASCE, SA3, 96 (1970), 653–663.

Pennypacker, Stanley P.; Sopper, William E.; and Kardos, Louis T. "Renovation of Wastewater Effluent by Irrigation of Forest Land," *Journal of the Water Pollution Control Federation*, 39 (1967), 285–296.

Perloff, Harvey S., and Wingo, Lowdon, Jr., eds. *Issues in Urban Economics* (Baltimore, Johns Hopkins Press, 1968).

Perrucci, Robert. "Engineering: Professional Servant of Power," *American Behavioral Scientist*, 14 (1971), 492–506.

Postlewait, John C., and Knudsen, Harry J. "Some Experiences in Land Acquisition for a Land Disposal System for Sewage Effluent," in *Recycling Municipal Sludges and Effluents on Land*, ed. Environmental Protection Agency, Proceedings of a Joint Conference on Recycling Municipal Sludges and Effluents on Land, Champaign, Illinois, July 9–13, 1973 (Washington, D.C., 1974).

Pound, Charles E.; Crites, Ronald W.; and Griffes, Douglas A. *Costs of Wastewater Treatment by Land Application*, Environmental Protection Agency, Office of Water Program Operations, Technical Report EPA-430/9-75-003 (Washington, D.C., 1975).

Rabinovitz, Francine F. *City Politics and Planning* (New York, Atherton, 1969).

Rados, David. "Selection and Evaluation of Alternatives in Repetitive Decision-Making," *Administrative Science Quarterly*, 17 (1972), 196–204.

Raffel, Marshall W. "Education for Health Services Administration," *American Journal of Public Health*, 60 (1970), 982–987.

Ralph Stone and Company, Inc. *Wastewater Reclamation: Socio-Economics, Technology, and Public Acceptance* (Springfield, Virginia, National Technical Information Service, 1974).

Riley, Matilda W., et al. "Socialization for the Middle and Later Years," in *Handbook of Socialization Theory and Research*, ed. D. A. Goslin (Chicago, Rand McNally, 1968).

Roe, Anne. *The Psychology of Occupations* (New York, John Wiley, 1956).

Rogers, Everett M. *Diffusion of Innovations* (New York, Free Press, 1962).

Rokeach, Milton. "Faith, Hope, and Bigotry," *Psychology Today*, 3 (April, 1970), 33–37, 58.

————. "The Role of Values in Public Opinion Research," *Public Opinion Quarterly*, 32 (1968–69), 547–559.

Romani, John. "The Administration of Public Health Services," in *Politics, Professionalism and the Environment*, ed. Lynton Caldwell, Environmental Studies, 3 (Bloomington, Ind., Institute of Public Administration, 1967), pp. 1–21.

Rose, John L. "Injection of Treated Waste Water into Aquifers," *Water and Wastes Engineering*, 5 (1968), 40–43.

Rosenthal, Donald B., and Crain, Robert L. "Structure and Values in Local Political Systems: The Case of Fluoridation Decisions," *Journal of Politics*, 28 (1966), 169–196.

Russell, Clifford W.; Arey, David G.; and Kates, Robert W. *Drought and Urban Water Supply* (Baltimore, Johns Hopkins Press, 1970).

*San Angelo Times*. March, 1971.

Sapolsky, Harvey M. "Science, Voters, and the Fluoridation Controversy," *Science*, 162 (October 25, 1967), 427–432.

Schmidt, Curtis J., Beardsley, R. F.; and Clements, E. V., III. "A Survey of Industrial Use of Municipal Wastewater," in *Complete Water Reuse: Industry's Opportunity*, ed. Lawrence K. Cecil, Proceedings of the National Conference on Complete Water Reuse, American Institute of Chemical Engineering, April, 1973 (New York, AICE, 1973).

————, and Clements, Ernest V., III. *Demonstrated Technology and Research Needs of Municipal Wastewater* (Cincinnati, Ohio, U.S. Environmental Protection Agency, 1975).

Schumpeter, Joseph H. *The Theory of Economic Development* (New York, Oxford University Press, 1961).

Sebastian, Frank P. "Wastewater Reclamation and Reuse," *Water and Wastes Engineering*, 7 (1970), 46–47.

Sewell, W. R. Derrick. "Environmental Perceptions and Attitudes of Engineers and Public Health Officials," *Environment and Behavior*, 3 (1971), 23–59.

————. "The Role of Attitudes of Engineers in Water Management," in *Attitudes towards Water: An Interdisciplinary Exploration*, ed. Fred L. Strodtbeck and Gilbert F. White (privately distributed, 1970).

————, et al., eds. *Forecasting the Demands for Water* (Ottawa, Canada, Department of Energy, Mines, and Resources, 1968).

————, and Burton, Ian, eds. *Perceptions and Attitudes in Resources Management* (Ottawa, Canada, Policy Research and Coordination Branch, Department of Energy, Mines, and Resources, 1971).

————, and Little, Brian R. "Specialists, Laymen and the Process of Environmental Appraisal," *Regional Studies*, 7, No. 2 (1973), 161–171.

Shuval, H. I., and Katzenelson, E. "The Detection of Enteric Virusus in the Water Environment," in *Water Pollution Microbiology*, ed. Ralph Mitchell (New York, Wiley-Interscience, 1972).

Simmons, John. "Economic Significance of Unaccounted-For Water," *Journal of the American Water Works Association*, 58 (1966), 639–641.

Simon, Herbert. "The Decision Maker as Innovator," in *Concepts and Issues in Administrative Behavior*, ed. Sidney Malich and Edward Van Niss (Englewood Cliffs, New Jersey, Prentice Hall, 1962).

————. "On the Concept of Organizational Goal," *Administrative Science Quarterly*, 9 (1964), 1–22.

Stander, G. J., and Funke, J. W. "Conservation of Water Reuse in South Africa," *Chemical Engineering Progress Symposium Series*, 63, No. 78 (January 1969), 3–14.

Stephan, David, and Robert B. Schaffer. "Wastewater Treatment and Renovation Status of Process Development," *Journal of the Water Pollution Control Federation*, 42 (1970), 399–410.

Stevens, Leonard A. "Every Drop Counts," *National Civic Review*, 56 (1967), 142–147, 155.

Strodtbeck, Fred L., and Gilbert F. White, eds. *Attitudes toward Water: An Interdisciplinary Exploration* (privately distributed, 1970).

Struening, Elmer L., and Lehmann, Stanley. "Authoritarian and Prejudicial Attitudes of University Faculty Members," in *The Engineer and the Social System*, ed. Robert Perrucci and Joel Gerstl (New York, Wiley, 1969).

Suchman, Edward A. "Health Attitudes and Behavior," *Archives of Environmental Health*, 20 (1970), 105–110.

Sullivan, Richard H.; Cohn, Morris M.; and Baxter, Samuel S. *Survey of Facilities Using Land Application of Wastewater* (Washington, D.C., U.S. Environmental Protection Agency, 1973).

Symons, George E. "That GAO Report," *Journal of the American Water Works Association*, 66 (1974), 275–276.

————. "Water Reuse—What Do We Mean?," *Water and Wastes Engineering*, 5 (1968), 40–43.

Symons, James M., et al. *National Organics Survey for Halogenated Organics in Drinking Water* (Cincinnati, EPA Water Supply Research Laboratory, 1975).

Tamir, O. "Administrative and Legal Aspects of Water in Israel," in *Water in Israel*, Part A, ed. Israel Ministry of Agriculture (Tel Aviv, 1973).

Tchobanoglous, George, and Eliassen, Rolf. "The Indirect Cycle of Water Reuse," *Water and Wastes Engineering*, 6 (1969), 35–41.

Thomas, Richard E. "Land Disposal II: An Overview of Treatment Methods,"

*Journal of the Water Pollution Control Federation*, 65 (1973), 1476–84.

Thompson, James, and McEwen, William. "Organizational Goals and Environment: Goal Setting as an Interaction Process," *American Sociological Review*, 23 (1958), 23–31.

United Nations, Statistical Office. *Statistical Yearbook 1972* (New York, 1973).

U.S. Army Corps of Engineers, Office of the Chief of Engineers. *Wastewater Management Program: Study Procedure* (Washington, D.C., 1972).

U.S. Bureau of the Census. *General Population Characteristics: Texas*, OC (1)-B45, Census of Population (Washington, D.C., 1970).

U.S. Bureau of the Census. *Statistical Abstract of the United States* (93rd ed., Washington, D.C., Government Printing Office, 1972).

U.S. Comptroller General of the United States. *Improved Federal and State Programs Needed to Insure the Purity and Safety of Drinking Water in the United States* (Washington, D.C., 1973).

U.S. Congress. *Federal Water Pollution Control Amendments of 1972*, Public Law 92-500, October 18, 1972, 92nd Cong. 82 Stat. 816, 33 USCA (1972).

U.S. Congress. *Water Quality Act of 1965*, Public Law 89-234, 89th Cong., 1st Sess. (1965).

U.S. Congress. *Water Resources Planning Act of 1965*, Public Law 85-90, 89th Cong., 1st Sess. (1965).

U.S. Congress, Senate Select Committee on National Water Resources, *Water Resources Activities in the United States* (Washington, D.C., Government Printing Office, 1960).

U.S. Department of Health, Education and Welfare. *Fluoridation Census, 1967* (Washington, D.C., Government Printing Office, 1968).

U.S. Department of Health, Education and Welfare. *A Strategy for a Liveable Environment* (Washington, D.C., Government Printing Office, 1967).

U.S. Department of Health, Education and Welfare, Public Health Service. *Public Health Service Drinking Water Standards: Revised 1962*, PHS Bulletins 956 (Washington, D.C., Government Printing Office, 1969).

U.S. Department of Health, Education and Welfare, Public Health Service. *Summary Report: The Advanced Waste Treatment Program* (Washington, D.C., 1965).

U.S. Department of Health, Education and Welfare, Public Health Service, Bureau of Water Hygiene. *Community Water Supply Study* (Washington, D.C., 1970).

U.S. Department of the Interior, Federal Water Pollution Control Administration. *The Economics of Clean Water* (3 vols., Washington, D.C., 1970).

U.S. Department of the Interior, Federal Water Pollution Control Administration. *Summary Report: Advanced Waste Treatment, July 1964–July 1967*. (Cincinnati, FWPCA, 1968).

U.S. Environmental Protection Agency. "Interim Primary Drinking Water Standards," *Federal Register*, 40, No. 51 (March 14, 1975), 11990–98.

U.S. Federal Water Resources Commission. *The Nation's Water Resources* (Washington, D.C., Government Printing Office, 1968).

U.S. National Water Commission. *Water Policies for the Future* (Washington, D.C., Government Printing Office, 1973).

U.S. Office of Saline Water. *Saline Water Conversion Summary Report, 1971–1972* (Washington, D.C., Government Printing Office, 1972).

U.S. Water Resources Council. *The Nation's Water Resources* (Washington, D.C., Government Printing Office, 1968).

U.S. Water Resources Council. "Principles and Standards for Planning Water and Related Land Resources," *Federal Register*, 38, No. 175 (September 10, 1973), 24778–869.

"Use of Reclaimed Wastewaters as a Public Water Supply Source," *Journal of the American Water Works Association*, 65, Pt. 2 (September, 1973), 64.

Van Burkalow, Anastasia. "The Geography of New York City's Water Supply: A Study of Interactions," *Geographical Review*, 49 (1959), 369–386.

Viessmann, Warren, Jr. "Developments in Waste Water Re-Use," *Public Works*, 96 (1965), 138–140.

Water Pollution Control Federation, Water Reuse Committee. "Policy Recommendations on Direct Reuse of Water" (Washington, 1971). (mimeographed)

Watkins, George Alfred. "Developing a 'Water Concern' Scale," *Journal of Environmental Education*, 5 (1974), 54–58.

———. "Public Attitudes on Drinking Reclaimed Wastewater." Paper presented at the American Water Works Association Meetings (1974). (mimeographed)

Wesner, G. M., and Baier, D. C. "Injection of Reclaimed Water into Confined Aquifers," *Journal of the American Water Works Association*, 62 (1970), 203–210.

White, Gilbert F. "Formation and Role of Public Attitudes," in *Environmental Quality in a Growing Economy*, ed. Henry Jarrett (Baltimore, Johns Hopkins Press, 1966), pp. 105–127.

———. *Natural Hazards* (New York, Oxford University Press, 1974).

———. *Strategies of American Water Management* (Ann Arbor, University of Michigan Press, 1969).

———. "Man's Alteration of the Water Cycle," in *Attitudes toward Water: An Interdisciplinary Exploration*, ed. Fred L. Strodtbeck and Gilbert F. White (privately distributed, 1970).

Whitford, Peter W. *Forecasting Demand for Urban Water Supply*, Report EEP-36 (Palo Alto, Calif., Stanford University, 1970).

Whyte, William F. *Money and Motivation* (New York, Harper, 1955).

Williams, Oliver, and Adrian, Charles. *Four Cities: A Study in Comparative Policy Making* (Philadelphia, University of Pennsylvania Press, 1963).

Wolfe, Harold W., and Esmond, Steven E. "Water Quality for Potable Reuse of Waste Water," *Water and Sewage Works*, 121 (1974), 48–54.

Wollman, Nathaniel. *The Value of Water in Alternative Uses* (Albuquerque, University of New Mexico Press, 1962).

————, and Bonem, Gilbert W. *The Outlook for Water* (Baltimore, Johns Hopkins Press, 1971).

World Health Organization. *Reuse of Effluents: Methods of Wastewater Treatment and Health Safeguards*, WHO Technical Reports, 517 (Geneva, 1973).

Yarden, D. *Drought Compensation Payments in Israel*, Natural Hazards Research Working Papers, 24 (Boulder, Institute of Behavioral Studies, University of Colorado, 1974).

York, C. Michael. "Opinions Relating to Fluoridation: Development of Measures," *Water Resources Bulletin*, 6 (1971), 920–924.

Zaltman, Gerald, and Lin, Nan. "On the Nature of Innovations," *American Behavioral Scientist*, 14 (1971), 651–674.

Zobler, L., et al. *Benefits from Integrated Water Management in Urban Areas—The Case of the New York Metropolitan Region* (Alexandria, Va., National Technical Information Service, 1969).

# Index